Clinical Ethics
Theory and Practice

Clinical Ethics: *Theory and Practice*, edited by *Barry Hoffmaster, Benjamin Freedman, and Gwen Fraser*, 1988

What Is a Person?, edited by *Michael F. Goodman*, 1988

Advocacy in Health Care, edited by *Joan H. Marks*, 1986

Which Babies Shall Live?, edited by *Thomas H. Murray and Arthur L. Caplan*, 1985

Feeling Good and Doing Better, edited by *Thomas H. Murray, Willard Gaylin, and Ruth Macklin*, 1984

Ethics and Animals, edited by *Harlan B. Miller and William H. Williams*, 1983

Profits and Professions, edited by *Wade L. Robison, Michael S. Pritchard, and Joseph Ellin*, 1983

Visions of Women, edited by *Linda A. Bell*, 1983

Medical Genetics Casebook, by *Colleen Clements*, 1982

Who Decides?, edited by *Nora K. Bell*, 1982

The Custom-Made Child?, edited by *Helen B. Holmes, Betty B. Hoskins, and Michael Gross*, 1981

Birth Contol and Controlling Birth, edited by *Helen B. Holmes, Betty B. Hoskins, and Michael Gross*, 1980

Medical Responsibility, edited by *Wade L. Robison and Michael S. Pritchard*, 1979

Contemporary Issues in Biomedical Ethics, edited by *John W. Davis, Barry Hoffmaster, and Sarah Shorten*, 1979.

THE LONDON FOUNDATION

Clinical Ethics

Theory and Practice

Edited by

Barry Hoffmaster

University of Western Ontario, London, Ontario

Benjamin Freedman

McGill University, Montreal

and Gwen Fraser

University of Western Ontario, London, Ontario

Humana Press • Clifton, New Jersey

Crescent Manor
PO Box 2148
Clifton, NJ 07015

Library of Congress Cataloging-in-Publication Data

Clinical Ethics: Theory and Practice/edited by Barry Hoffmaster,
    Benjamin Freedman and Gwen Fraser.
    p. cm. — (Contemporary issues in biomedicine, ethics, and society)
    Based on lectures delivered at a conference on the Nature and Teaching of Applied
Ethics in Medicine held in London, Ontario,
Canada, April 22–25, 1986 and sponsored by the Westminster Institute for Ethics and
Human Values and the Department of Philosophy, University of Western Ontario.
    Includes index.

    ISBN-13: 978-1-4612-8221-1   e-ISBN-13: 978-1-4612-3708-2
    DOI: 10.1007/978-1-4612-3708-2

    1. Medical Ethics—Congresses. I. Hoffmaster, C. Barry. II. Freedman,
Benjamin. III. Fraser, Gwen. IV. Westminster Institute for Ethics and Human Values. V.
University of Western Ontario. Dept. of Philosophy. VI. Series.

    [DNLM: 1. Ethics, Medical —Congresses. W 50 N285 1986] R724.N38 1988
174'.2—dc19
DNLM/DLC
for Library of Congress   88–13579
                                                 CIP

Preface

There is the world of ideas and the world of practice; the French are often for suppressing the one and the English the other; but neither is to be suppressed.
—Matthew Arnold
The Function of Criticism
at the Present Time

From its inception, bioethics has confronted the need to reconcile theory and practice. At first the confrontation was purely intellectual, as writers on ethical theory (within philosophy, theology, or other humanistic disciplines) turned their attention to topics from the world of medical practice. Recently the confrontation has grown more intense. The appointment of clinical ethicists in hospitals and other health-care settings is an accelerating trend in North America. Concomitantly, those institutions involved in training people in clinical ethics have added organized exposure to the world of practice, in the form of placement requirements, to the normal academic course load. In common with other disciplines, bioethics has begun to see clinical training as a condition of didactic theory and apprenticeship.

The mismatch between theory and practice in bioethics may be too deep to yield to this "both/and" approach—as its persistence as an interdisciplinary pursuit has resisted the

v

opposite temptation of "either/or." The theoretical side of bioethics requires leisurely examination, firm information, sharp distinctions, and a well-defined problem. When a bioethical issue arises in practice, however, none of these conditions is likely to be obtained. In a clinical setting, questions do not arrive neatly labeled. The empirical background, including diagnosis and, especially, prognosis, is often contentious. And waiting for ideal circumstances for decision making will commonly result in a decision by default. Using academic modes of analysis on medical practice begins to resemble using a surgical laser to clear underbrush.

This volume examines the relationship between academic approaches to bioethics and the clinical context, in both theory and practice. The papers are based on and developed from lectures delivered at a conference on The Nature and Teaching of Applied Ethics in Medicine held in London, Ontario, Canada, April 22–25, 1986; the commentaries and rejoinders were submitted subsequently by participants who came from across North America, bringing to the conference an exciting variety of viewpoints and extensive and diverse experience.

The conference was organized by the Westminster Institute for Ethics and Human Values and by the Department of Philosophy, University of Western Ontario (UWO), both in London. Financial support was provided by the Strategic Grants Division (Human Context of Science and Technology) of the Social Sciences and Humanities Research Council of Canada (Grant Number 499-85-2002). Further financial support was granted by the Westminster Institute; the Department of Philosophy, UWO; and the London Foundation, which helped us make this material available to a wider audience. For their support of the conference, we gratefully acknowledge Abbyann Lynch, Ausonio Marras, W. Lockwood Miller and the Board of the Westminster Institute, and

Charles Rand. Françoise Baylis, Rob Butcher, Greg Del-Bigio, John Maher, and Bob Wright assisted in conference arrangements. Janet Baldock and Maxine Abrams facilitated the typing, correspondence, grant administration, and communication needed for the conference. Above all, Linda Nicholas' organizational abilities and commitment to the discipline and the conference were almost single-handedly responsible for turning a good idea into an excellent conference.

It is fitting that a conference dealing with theory and practice should have been held in Canada, a nation comprised of the theoretical French and the empirical English, which has been trying since its inception to reconcile the two. We hope that these papers will further the debate.

Barry Hoffmaster
Benjamin Freedman
Gwen Fraser

Contents

Contributors

Terrence F. Ackerman • *Department of Human Values and Ethics, College of Medicine, The University of Tennessee, Memphis, Tennessee*

Robert Baker • *Department of Philosophy, Union College, Schenectady, New York*

Françoise Baylis • *Department of Philosophy, The University of Western Ontario, London, Ontario*

Howard Brody • *Medical Humanities Program, Michigan State University, East Lansing, Michigan*

Arthur L. Caplan • *Center for Biomedical Ethics, University of Minnesota, Minneapolis, Minnesota*

K. Danner Clouser • *Department of Humanities, The Pennsylvania State University, College of Medicine, The Milton S. Hershey Medical Center, Hershey, Pennsylvania*

Benjamin Freedman • *Centre for Medicine, Ethics and Law, McGill University, Montreal, Quebec*

Barry Hoffmaster • *Department of Philosophy,
The University of Western Ontario, London, Ontario*

Ruth Macklin • *Department of Epidemiology and Social
Medicine, Albert Einstein College of Medicine
of Yeshiva University, Bronx, New York*

Robert M. Veatch • *Kennedy Institute of Ethics,
Georgetown University, Washington, DC*

Introduction

This volume does not deal with substantive issues in bioethics; rather, it is an attempt to come to terms with the nature of the discipline. Bioethics is in its adolescence, a propitious time to look at where it has been and to try to determine where it is going. As a field of study and research, as well as practice, bioethics has been a precocious child, developing rapidly and insinuating its way into medicine with fewer difficulties than some of the medical subspecialties have experienced. The measure of its success is that it has become accepted and, some might even say, useful. At the same time, however, it is nursing among its practitioners a potentially dangerous schism that results from two fundamentally different answers to the question, "What *is* applied ethics in medicine?"

There are, on the one hand, those who claim that the practice of bioethics involves the application of a favored moral theory to the facts of a case to arrive at a decision about how the case should be handled. Others who utilize this general approach adopt a less restrictive position that allows them to choose, among the set of rival moral theories, the one most suited to the case at hand and then apply that theory to the facts to reach a decision. Or they develop a fragment of an ethical theory, for example, one concerned solely with consent, to apply to aspects of selected problems. The alternative position claims that, whatever bioethics is, it is not applying traditional moral theories to the facts of cases to generate decisions that will be helpful to clinical practi-

1

tioners.   According to this view, moral principles are too general and abstract to yield determinate conclusions about specific cases. Decisions depend upon the morally relevant particularities of cases, which can be discovered only by examining the cases themselves.

Given this basic conceptual opposition, one can understand two major problems that infect bioethics. The first arises in the clinical setting. An increasing number of health care institutions in both the US and Canada are utilizing the services of clinical ethicists. These persons, often philosophers but from other academic disciplines or clerical backgrounds as well, consult with medical staff or a hospital policy making body. Consultations with staff members are usually for the purpose of clarifying or trying to resolve difficult cases wherein the issues are moral as well as clinical. Consultations with hospital administrators are for the purpose of establishing policies in regard to identifiable classes of problematic cases. The theoretic bent of the ethics consultant, whether principles-oriented or case-oriented, can profoundly affect this decision and policy making, however. The two approaches not only could yield different solutions to cases (resulting, perhaps, in different policies at the administrative level) but also could portray bioethics as two fundamentally different endeavors. Neither outcome would enhance the credibility of the discipline.

The second problem is pedagogical: How is bioethics taught and how should it be taught? Those who are principles-oriented maintain that students of bioethics need to be taught traditional moral theory, and many who adopt this approach disavow the necessity for clinical experience in an applied ethics program. On the other hand, those who are case-oriented view the teaching of traditional moral theory as a minor, perhaps even dispensable, part of a program in bioethics and insist that clinical experience is essential in the training of those who would practice as clinical ethicists. How one views the nature of bioethics also profoundly influences how it is taught.

The purpose of the conference that spawned this book was to bring together well-known proponents of both conceptions of bioethics to explore the dichotomy between theory and practice. The venture was an attempt to gather the opposing players in a common setting where an honest, spirited exchange of views could occur. Those who have contributed to this volume teach and write in bioethics. Most have also served as clinical ethics consultants. Thus, their papers are critical reflections on both their theoretical and practical experiences in the discipline.

An ancillary reason for holding the conference and publishing this collection is the attempt to foster an exchange between the Canadian and American centers with the most experience in clinical ethics. There is a need to identify and compare teaching strategies, whether the bioethics course is taught in a philosophy department or a medical school. The importance of this objective is demonstrated by the increasing number of bioethics courses listed in the Syllabus Exchange of the Kennedy Institute for Ethics. Unfortunately, although formal teaching programs are developing rapidly in universities, colleges, and professional schools, there has been little effort to assess their scope or worth. Any such assessment, however, presupposes a view about the nature of the discipline.

The sequence of the papers does not reflect the order in which they were delivered at the conference. Rather, it is intended to capture the contrast between a principles-based approach and a case-based approach. At the conference, Robert Veatch and Arthur Caplan, the keynote speakers, were charged with setting the framework for the papers and discussions that followed.

The principles-based approach is represented by Veatch and Macklin. Veatch outlines alternative roles that a clinical ethicist might assume, but the performance of these roles is constrained by the demands of traditional moral theory. It is also constrained by Veatch's insistence that ethical decisions be based on the patient's value system. In his commentary on Veatch's paper, Baker challenges the assumptions of Veatch's position and defends a more expansive role for a clinical ethicist.

Macklin addresses her remarks to skeptics who doubt that bioethics is useful in resolving moral problems. She argues that a moral philosopher can at least facilitate agreement on the morally relevant facts of a case. Further, she argues that, though an ethicist may not resolve cases, the ethicist can clarify issues, and the parties to a dispute are further ahead when they understand the nature of their disagreement.

The case-based approach is represented, in varying degrees, by Caplan, Ackerman, Brody, Clouser, and Hoffmaster, all of whom believe that there is a great deal more to the practice of bioethics than the application of moral principles to facts. Caplan discusses the frequently denigrated notions of moral expert and moral expertise. He argues that the concepts of expert and expertise are accepted in a variety of contexts and replies to objections to importing these notions into the moral sphere. Furthermore, he claims that moral philosophers have a role to play in clinical situations when expertise in ethics is required. Baylis analyzes the expert/expertise distinction further, and though she agrees with Caplan's characterization of moral expertise, she takes issue with his description of moral experts.

Brody believes that ethical decision making in clinical contexts can be understood in terms of the metaphor of conversation. He maintains that by adopting a procedural rule that prevents changing the subject when discussants reach difficult issues, a decision that is acceptable to all concerned parties can be reached. Ackerman takes a similar approach in proposing a decision-making process designed to elicit what he calls "a shared commitment to cherished states of affairs." Hoffmaster argues that the problems of applying moral theories to real-life cases are more serious than generally conceded. Making minor adjustments between theory and practice will not work, so, in his view, the principles-based approach should be jettisoned. Clouser's view is less extreme. He acknowledges the gap between ethical theory and applied ethics, but suggests that it can be bridged by regarding the progression from abstract theory to practical application as a continuum. He claims that applied ethics may inform ethical theory to a greater or lesser extent, depending upon how far one is prepared to extend the continuum toward the practical end.

The one paper in this volume that differs from the others is Freedman's. He argues for the formulation and adoption of a Code of Ethics for ethicists in clinical settings. Professional codes are necessary in any self-regulating profession, and ethicists, if they are to consider themselves members of a profession, ought not to be an exception, even if they are "experts" in ethics.

Whether this collection of papers has begun to reconcile divergent views about the nature of bioethics is left to the reader to decide. At the very least, however, these papers confront the fundamental theoretical problems facing bioethics as a growing discipline. The exchanges initiated here should provide a basis for continuing discussion among members of the profession.

Gwen Fraser

Clinical Ethics, Applied Ethics, and Theory

Robert M. Veatch

Introduction

The past fifteen years has seen a remarkable increase in interest in what is often called applied ethics. People trained in philosophical and religious ethics and those working in many other disciplines have fought to break loose from more purely theoretical investigations in metaethics and normative ethics that are not directly tied to moral problems that need solving. The cry has been for ethics in a real-life context, where the tools of ethics are used to solve, or at least clarify, dilemmas real persons are facing.

Some, but not all, of that applied ethics has been clinical. Tom Beauchamp defines applied ethics as "the use of philosophical theory and methods to analyze fundamentally moral problems in the professions, technology, public policy, and the like."[1] If applied ethics can be understood as ethics applied to real decisions made by real people, clinical ethics is narrower. Clinical ethics, as I am using the term, is always applied ethics, but it is restricted in at least two ways. First, clinical ethics ought to have something to do with "clinics." These need not be health care clinics (including hospitals and doctors' offices); they might be legal clinics or any other setting where people meet to interact over practical problem solving. For our purposes I shall limit the term "clinical ethics" to applied ethics involving interactions between professionals and laypersons excluding the large realm of applied ethics having to do with broader public policy matters as well as practical problem solving done by individuals without the benefit of outside consultants. Although I shall focus on professional/lay relationships, I believe nothing is lost if what is said here is applied to all applied ethics involving experts offering themselves to clients for a fee (such as repair persons, commissioned artists, or tour guides).

Second, I shall narrow the term clinical ethics even further. I shall limit it to ethical deliberations that take place close to the decision-making interactions (such as on a hospital floor or in a physician's office). Clearly some applied ethics dealing with lay/professional relationships is much further removed. Scholars trying to understand the history and meaning of the principle of confidentiality by doing comparative historical study are not doing clinical ethics by my definition, even if they hope that their readers will directly apply their conclusions regarding medical confidentiality. Clinical ethics, as I am using the term, is applied ethics at the scene of professional/lay relationships: for practical purposes in the clinic in its most generic sense.

The Relation of Theory to Application

There has been considerable controversy over the relation between ethical theory and applied ethics. For some time, many argued that applied ethics could be nothing more than the application of a pre-existing theoretical framework to a content area posing problems that need solving. Ethical theory provides a framework for analyzing moral dilemmas. Ethical theory often provides, as well, a set of moral norms, principles, or rules that yield a systematic approach or even solution to moral dilemmas. The alternatives to an ethical theory are an intuition, gut feeling, appeals to authority, or just blatant inconsistency. It is hard to imagine doing ethics without theory—explicit or implicit.

It is increasingly recognized, however, that no sharp distinction can be made between ethical theory and applied ethics. Moreover, those who spend their time working in quite specific areas of ethical analysis and problem solving such as medicine may have a great deal to offer to those who spend their time doing more abstract theory. Applied ethics provides the "considered moral judgments" that can be the data by which theory is tested and refined. Applied ethics provides an understanding of what ethical problems need to be addressed, of where the holes are in a theoretical framework. It may, in fact, suggest new theoretical options that would never occur to those working at levels detached from specific content.

For example, Hippocratic medical ethics offers a unique blend of consequentialist normative ethics ("benefit the patient according to your ability and judgment") with a radical truncation of traditional consequentialism. Traditional utilitarianism is committed to considering aggregate net benefits for all those potentially affected by an action. By contrast, Hippocratism considers benefits and harms, but only those affecting the patient.[2] As such it is really not taking only consequences into account. The person to whom the consequences accrue seems to be essential. It is as if for Hippocratists one, and only one, formalist consideration is allowed into play: the constraint that the only consequences that are allowed to count are those to the patient. No other major ethical theory of which I am aware makes such a move.[3]

This should be provocative to theorists. Why is it that no normative theory outside medicine makes such moves?

Should normative theory be adjusted to take this possibility into account; or, should the fact that no other ethical system encompasses such a strange mix of consequentialism with a single formalist constraint be taken as evidence that Hippocratism is mistaken?

Analysis of applied ethics can enrich ethical theory. Since clinical ethics is a kind of applied ethics, there is the possibility that clinical ethics can offer this to theorists. Moreover, since, among applied ethics, clinical ethics is as close as possible to actual moral decisions, those who are convinced of the importance of linking theory and concrete decision making should be particularly intrigued with clinical ethics.

Three Roles in Clinical Ethics

Thus for one who takes applied ethics seriously, both as a contribution to ethical theory and as a way of making a real difference in concrete moral decision making, clinical ethics has real possibilities. Beyond the possibility that clinical ethicists can have significant impact on ethical theory, at least three separate tasks are available for ethicists in a clinical setting. Unfortunately, not all of these tasks are viewed as justified by everyone, and probably they are, in some settings, incompatible.

The Ethicist as Analyst

The most conspicuous and easily defensible role for the ethicist in a clinical setting is that of analyst. Ethicists ought to be trained to know the ethical geography. They ought to be able to help identify the range of plausible positions on a given clinical problem and to recognize the kinds of arguments that would be appropriate to defend these positions. Thus, for example, if two medical residents are engaged in a heated argument on a hospital floor about whether to disclose a terminal malignancy to a depressed patient one might offer many arguments, based upon consequences, defending a nondisclosure policy. Should the other resident have only an intuition that it would be wrong to withhold the information, an ethicist trained in elementary theory should be able to help discover whether these residents are merely two consequentialists who disagree about the medical and physiological consequences of disclosure, or whether the resident reluctant to withhold is committed to some more Kantian duty of truth telling.[4]

Knowing this will make a significant difference to how they resolve their conflict. If they merely disagree about consequences, then they will need to develop some way of resolving empirical disagreements in order to resolve the conflict. They might search the psychological literature, for example, to attempt to discover the consequences of disclosing a critical diagnosis. If they are really disagreeing over ethical theory, however, then no amount of searching the psychological literature will help. Even if the two are really consequentialists engaged in an empirical disagreement over what will benefit the patient, the ethical analyst might be able to reveal that others in the action system—the patient, the nursing staff, the administration, or those who prosecute for failure to get adequately informed consent— may have other ethical norms not shared by the two in the dispute.

Such analysis can help reveal that it is literally impossible to determine what constitutes good medical practice until some critical ethical controversies are resolved. Often the ethicist in the role of analyst will feel compelled to adopt a position of neutrality. It is argued that the ability to recognize the nature and range of arguments in no way implies expertise in deciding which of the two plausible normative

positions is correct. Still less does it give one skill in resolving complex empirical disagreements. Being trained in
medical science does not in itself give one any authority to
make choices among alternative medical interventions, because deciding among alternative medical interventions
depends, in part, on ethical and other value considerations.
So, likewise, being trained in the science of ethical analysis
does not give one authority to make choices among alternative moral positions at the level of concrete ethical choice.
Different moral traditions and different systems of empirical belief and value lead to different decisions. Different
competent ethicists appear to stand in about as wide a range
of these traditions as other persons in society. Being skilled
in the analysis of ethical positions does not imply expertise
in making ethical choices.

The Ethicist as Advisor

Sometimes the ethicist is cast in a radically different
role, that of moral advisor. If concrete choices are likely to
be influenced (justifiably) by different systems of belief and
value, then it makes sense that the expert in analysis might
not also be an expert in particular traditions. On the other
hand, some of those traditions do recognize authorities in
the values and normative judgments made within their traditions. Priests have special teaching and interpretative
authority in many religious traditions. Mystics and pro-
phets are respected by their followers. Some secular com-
munities recognize a cadre of wise or charismatic persons
whose judgments are respected on moral matters.

In some clinical settings it might be perfectly appropriate to have such persons in residence to offer advice (or even
definitive authoritative judgments) about moral questions
that arise. A rabbinical council at a local Jewish hospital
might be assumed by all involved to have substantial moral

authority. A family priest might do far more than simply analyze moral alternatives for a devout Catholic patient from his parish.

Perhaps the finest example of this role of ethicist as advisor is Father Trapasso, priest of the Karen Quinlan family. At the time of Karen Quinlan's tragic accident, the family had not studied thoroughly the ethical issues related to termination of treatment from the perspective of their own Catholic heritage. They turned to their parish priest who, in some ways, took on a role of ethics teacher and advisor.[5] He surely did some analysis. He introduced important distinctions from within their shared tradition as well as others, distinctions between active killing and letting die, between withdrawing and active killing, and, most importantly, between ordinary and extraordinary means. He went far beyond analyzing, however. In fact, he may not have been particularly trained to provide a general analysis of the moral options, especially those from outside his own heritage. His role extended to that of advisor. It is my understanding that he made substantive moral recommendations to the effect that it is not morally obligatory to continue treatment of patients in a vegetative state. To the extent that the person in the clinical ethics role is legitimated as morally authoritative by the one getting the advice, such pronouncements are appropriate and justified.

Ethicists might be called upon in this fashion to serve as moral advisors not only to patients, but to physicians or to institutional structures (such as the Board of Trustees). Many ethicists, trained in secular, academic, and philosophical programs, might reject such a role; or, more likely, they might never be recognized by physicians or patients as legitimately occupying this role. But some people referred to as clinical ethicists, especially within sectarian institutions, may be expected to play the role of advisor.

These two roles of analyst and advisor create potential problems. A clinical ethicist hired by a secular teaching hospital to do primarily analytical tasks might be treated by some physicians as capable of giving authoritative moral advice. A personal example illustrates the point. When on the floors of Columbia Presbyterian Hospital as part of my work with the Columbia College of Physicians and Surgeons Medical Ethics Program in the early 1970s, I was asked to give an "ethics" consult for the nephrology service by a resident. It became very clear that he expected a definitive moral decision from me about whether to dialyze a critically ill patient who had little chance of success from the treatment.

My response then and in similar situations depends upon my determination of how confused the physician is about my ability to make definitive, *ex cathedra*, moral pronouncements. If he believes I have special powers of this sort, I will refuse to reveal my conclusions. If I fear he will immediately disconnect a patient from life support merely because I, as one ethicist, believe, based on my particular beliefs and values, that it is acceptable, then I have a duty as a resident ethicist to refuse to reveal those preferences. However, if the clinician has an appropriately low opinion of my capacity to make definitive moral judgments (as opposed to providing analyses), then I feel free to unload on him or her and to argue with all the force I can muster. I am comforted in knowing that the clinician, who is, in effect, the ethicist's patient, the ethicist's client, will not be influenced beyond what the argument will bear.

I believe the clinician has exactly the same obligation to the patient. If the clinician believes that a patient will accept medical treatment simply because the clinician favors it based on his or her beliefs and values, then the clinician should refuse to reveal that preference. It is impossible to know whether to treat without drawing on some values. There is no reason why the patient should use the physician's values in deciding whether a treatment is appropriate. On the other hand, the clinician who is fortunate enough to have a patient who is adequately skeptical of his or her clinical opinions should feel free to argue with the patient for the therapeutic regimen the clinician prefers. Such a patient would likely be influenced only to the extent that the physician's values are plausible.

The Ethicist as Adversary

This suggests a third possible role for the clinical ethicist: that of adversary. It is natural for intelligent, thinking persons to enjoy arguing for positions they hold, or at least consider, without feeling that either they themselves or others will consider these positions definitive and authoritative. Ethicists in a clinical setting have at least two things going for them that recommend this ambiguous role. First, as people trained in ethical analysis, they should be able to identify positions not normally represented in clinical decision making processes. Second, since they come from cultures and academic disciplines with traditions far different from those of clinical medicine, they often will be inclined to different theoretical and casuistic moral positions. They are often added to hospital ethics committees and Institutional Review Boards (IRBs) as much for this role as for the capacity to analyze or to advise.

It should be clear that this is not technically a role for which they have been trained. They may have been trained as analysts or as authoritative interpreters of a particular tradition or both, but other educated persons in the community who have the courage to challenge common wisdom could play the adversarial role at least as well. A local lawyer, a humanist from outside the discipline of ethics, or an adequately aggressive business person may do just as well. A clinical ethicist in residence in a hospital often may play this role with physicians, patients, and administrators.[6]

The most troubling thing is that playing any one of these roles well may exclude playing the other roles. The ethicist in the role of advisor may be diminished in the capacity to analyze dispassionately. The feisty adversary who does a good job at arguing unpopular positions in a clinical context may lose respect as an analyst. Those strongly committed to one tradition and good at advising what that tradition says may not be skilled enough or dispassionate enough to be analysts. The rabbinical scholar or the advocate of libertarianism may be examples.

Problems With Clinical Ethics

This suggests that there are several important roles that a clinical ethicist might play in a hospital setting, but that there may also be tensions inherent in the job as well. In addition to the tensions among the different roles that the ethicist might be expected to take on, there are other potential problems with the notion of systematic ethics being done in a clinical setting. These deserve our attention.

Identifying the Primary Decision Maker

Clinical ethics involves real-life decisions and is particularly concerned with real-life decisions in a clinic, that is, in a context where a professional such as a health care professional is interacting with a layperson. If that is the case, then a special problem exists for ethicists who function in a clinical setting. They have to make certain that they are doing the analysis, giving their advice, or serving as an adversary for the appropriate decision maker.

It is commonplace for those trained in ethics who are working in a hospital setting to speak of providing these functions for the clinician "on the firing line," for the physicians, nurses, and possibly the administrators who are thought to have to make actual clinical decisions. That model, however, buys into the idea that the physician, nurse, or administrator is the one who is faced with the clinical ethical choices. That idea is much more controversial than it may appear. It lines the ethicist up on the side of the professional and assumes that it is the professional, rather than the layperson, who is the decisive decision maker.

Increasingly it is recognized that this is a serious mistake. Medical decisions necessarily involve ethical or other value components. In every single medical choice one course of action is preferred over another. This is true not only for stereotypically ethical issues such as euthanasia, abortion, and allocating scarce resources. It is equally true for every routine choice that is made in the clinical setting (or anywhere else). To say that a drug is "medically indicated" means nothing more than it is preferred over other alternatives on the basis of the value system of the one expressing the judgment. It is not in any way a matter of scientific "fact" (unless one is willing to talk about value judgments being a particular kind of factual statement).

If this is so, then every clinical choice must be preferred on the basis of some value set. Insofar as decisions are clinical, that is, ones that affect the layperson, the appropriate value set, at least to a large extent, is that of the layperson. There is no evidence whatsoever that professionals in a clinical field such as medicine have an expertise at choosing the values upon which clinical decisions must be based. Even if there were evidence that the clinician were better at choosing a value system than the patient, any person committed to the principle of autonomy would still prefer that, insofar as the decision affects only the patient, the patient's value system be used as the basis for the clinical choice. Insofar as the choice is social, the value systems of others are clearly relevant, but there is no basis for assuming that the other value system that should be decisive should be that of the clinician or of the collectivity of clinicians. In fact a good argument can be made that insofar as a value choice is social, the clinician should be exempted from any responsibility, the clinician's responsibility being to the patient rather than to some social unit.[7]

The implications of the above claims are far more radical than most people realize. If every clinical decision involves a value set and if the appropriate value set is not normally that of the clinician, then the ethicist in the clinical setting is making an error if he or she offers analysis, advice, or adversarial opposition primarily for the clinician.

It is true that it is impossible for any professional—ethicist or physician—to interact with a client without making certain choices along the way. Those choices, however, are, or should be, severely limited. There are only three kinds of choices of which I am aware that clinicians ought to be making in a liberal society. First, they should retain the right to make decisions based on conscience to withdraw from a clinical relationship when the client's value system leads to a decision that is so morally repugnant that the clinician cannot in good conscience continue; provided, within certain limits, that other professionals are available to step in and provide care.

Second, clinicians must make inevitable decisions that can be called "initiation decisions." In counseling with clients they must select from a long list of possible trial diagnoses and therapeutic interventions in order to know what choices to give the client. For example, I once studied the impact of physician value systems on whether they chose to initiate conversations with obstetrical patients about contraception during premarital examinations.[8] The question was whether, in cases where the patient did not initiate conversation, the physician should bring up the subject of contraception. The intriguing thing about this clinical choice is that it is logically impossible to ask what the patient's preference is. To do so would be to initiate the conversation and thus answer the question being asked. Similar "initiation decisions" must be made by clinicians in deciding whether to present very unlikely therapeutic alternatives or to mention during a consent process side effects that the patient is very unlikely to want to know about. Clinicians are necessarily decision makers in these situations.

Third, clinicians are appropriate decision makers when patients autonomously yield decision-making authority to the clinician as a surrogate. Although there might be moral debate over whether patients ought to yield such authority to clinicians, sometimes they do, and most people acknowledge that they are within their rights in doing so.

In these three situations clinicians are necessarily decision makers who must select a set of values upon which to make their choices. In such circumstances a clinical ethicist would appropriately provide analysis, advice, or adversarial opposition for the clinician. Those are very restricted circumstances, however. Otherwise, the clinical ethicist's service to the clinician should be restricted to pointing out that it is the patient who is the appropriate decision maker. Clearly, if one is committed to respecting the patient and the patient's value system, most of the critical choices ought to be with the patient.

It is helpful to distinguish between what could be called primary and secondary decision makers. In every clinical situation one person, normally the patient or the surrogate for the incompetent patient, is properly viewed as a primary decision maker, as the one whose value system ought to govern the primary decisions being made. Others, however, are what could be called secondary decision makers. They must make certain tangential decisions such as whether their consciences allow them to participate, and what special professional knowledge might be of interest to the primary decision maker. Clinicians are secondary decision makers in this sense. This is not to say that their decisions are of secondary moral importance. A decision of Elizabeth Bouvia's physicians to refuse to participate in her chosen plan of starving herself to death while under health professionals' care is an extremely important moral choice.[9] Analytically, however, it is secondary to Ms. Bouvia's choice to reject medically administered feeding in a hospital setting and to request instead administration of support while she dies of starvation.

What does this mean for clinical ethicists? If they view themselves as servants of clinicians, they have taken a controversial stand on the issue of who ought to be the primary decision makers. Clinical ethicists, it would seem, ought to be making themselves available primarily to the primary decision makers. They also have some responsibility to the secondary decision makers when they face difficult choices about conscientious objection, initiation decisions, and so on; but those are surely subordinate to the critical choices that the primary decision makers must make. If those choices are to be made by patients, then clinical ethicists ought to have the patient as the primary client. The fact that the ethicist is on the clinician's turf, is paid by the health professional system, and gradually develops identification with clinical professionals, all cast doubt on the legitimacy of the clinical ethicist's role.

Dealing with Critical Decisions Outside the Clinic

There is a second problem in placing the clinical ethicist on the staff of the clinic or otherwise connecting the ethicist's work with the clinic. The model seems to be one that implies that critical decisions are made primarily in the clinic. At least in the clinical setting (defined as one where applied ethical and other value choices are made in the context of lay–professional relationships), the layperson has the advantage of the other advisor who has experience in dealing with the issues under discussion. In many other settings he or she does not.

In fact, many, perhaps most, critical value choices involving applied ethical decision making take place outside the clinical setting. Patients decide whether or not to seek professional counsel. They decide whether to take their child with a high fever to the hospital emergency room. They decide whether to self-medicate. They choose which of many professional advisory systems they should use. They decide whether to make certain practices illegal. It is fair to say that many critical choices are made without interaction with a health professional.

Moreover, even when they are interacting with a clinical professional, it is often more appropriate for them to get their ethical analysis, advice, or adversarial opposition from someone other than a person associated with the professional's framework. A Catholic patient trying to decide whether it is ethical to refuse life-prolonging nasogastric feeding may more appropriately turn to the priest than to the ethicist in the payroll of the hospital to which the patient has been taken.

Even for analysis of the issues, the ethicist on the professional's turf may not be the right person to whom the layperson ought to turn. Analysis of options is not a value-free enterprise, but is contingent upon the analyst's world view. The in-house, all-purpose analyst might not be able to do the job for the patient even if the analyst is sincerely committed to fair, "neutral" analysis. The ethicist trained in secular analytical philosophy may not even consider an option that would be crucial to a patient's own tradition. He might not think of discussing the problems of dietary choices with the patient who was, unbeknownst to him, a Seventh Day Adventist.

Many applied ethical problems appropriately needing the applied ethicist's assistance arise outside the clinic. Even for those that arise inside the clinic, the appropriate ethicist may not be the one on the clinic payroll and sharing the clinic's perspective. In certain cases, patients choose their hospitals precisely because they like the biases and value commitments of the institution. The Orthodox Jewish patient may purposely choose an Orthodox Jewish hospital. If he gets an ethicist imbued with an Orthodox perspective, no problem arises. But more often than not, this purposeful pairing will not take place. In those cases, the house ethicist may not be the right one. For the many ethical decisions that must be made outside the clinic, the house ethicist cannot help at all.

Overidentifying with the Health Care Professions

A final potential problem for clinic-based applied ethicists is the possibility of overidentification with the clinician's perspective. Any clinical ethicist who is hired by, or even tolerated by, a clinic has passed an initial screening. Placing an anti-abortion ethicist in the staff as clinical ethicist in an abortion clinic or hiring an opponent of external gamete manipulation as house ethicist for an in vitro fertilization clinic invites chaos. Even though they might be very good at challenging the assumptions of the primary and secondary decision makers, they probably would not be hired and could not survive if they were. The ethicists given clinical responsibility are a subset of those who might be chosen.

More subtly, clinical ethicists are likely to have world views like those of clinical professionals. They are at least likely to be case oriented. They have probably taken positions in ethical theory that are more akin to the professional staff than others. For example, they might be committed to situationalism rather than a rules of practice view.[10] They might be more likely to be consequentionalists than deontologists. Some, myself included, were originally trained as health professionals. We continue to carry licenses to work as clinicians. Our thought processes were shaped by the clinical sciences. We share the clinical mentality. (My ethics is far more empirical than that of many of my colleagues.) Although that may make it easier for some who work in clinical ethics to communicate with and understand health professionals and may make them more acceptable to health professionals, it can also seriously compromise their ability to play the roles they ought to be playing.

Insofar as they perform analytical tasks, they approach the analysis from the clinical perspective. Some may not even have exposure to the full range of philosophical and theological disciplines necessary to do their jobs well.[11] Insofar as they see themselves as analysts and advisors and adversaries, they may end up analyzing and advising and providing adversarial opposition for the health professionals who share their world view and whom they begin to view as their colleagues and friends.

The result is a kind of paradox. It seems that to function well as a clinical ethicist one ought to know the subject matter and working style of the relevant clinicians very well. Yet the more one identifies with the clinical material and mentality, the less well one can perform any of the tasks to which a clinical ethicist is appropriately assigned.

Conclusion

The job of clinical ethicist is an intriguing one. It has gotten greater attention as applied ethics gains greater prominence and clinical problems become more and more conspicuously ethical. The functions of analying, advising, and serving as adversary are all important. They need to be done. But they need to be done in ways that avoid losing sight of the fact that clinical decisions necessarily involve ethical and other value choices and that those choices, insofar as they are clinical, should involve primarily the patient or surrogate's value systems and only secondarily those of the professional clinician. Whether the clinical ethicist on the payroll of or receiving the sanction of the clinic and sharing at least to some degree the world view, thought processes, experience, and friendship of the clinical professionals can adequately do the job remains to be seen.

References

[1]Tom Beauchamp (1984) "On Eliminating the Distinction Between Applied Ethics and Ethical Theory," *The Monist* **67,** 514–531.

[2]Ludwig Edelstein (1967) "The Hippocratic Oath: Text, Translation, and Interpretation, "*Ancient Medicine: Selected Papers of Ludwig Edelstein,*" Owesi Temkin and C. Lilian Temkin, eds., Johns Hopkins Press, Baltimore, MD, pp. 3–64.

[3]Robert M. Veatch (1981) *A Theory of Medical Ethics,* Basic Books, New York, NY.

[4]Immanuel Kant, "On the Supposed Right to Tell Lies from Benevolent Motives," translated by Thomas Kingsmill Abbott and reprinted in Kant's *Critique of Practical Reason and Other Works on the Theory of Ethics* (1909 [1797], Longmans, London), pp. 361–365. For another example, see my analysis of a case involving two residents in a dispute over how aggressively to medicate for pain a patient with a broken leg,

in *Case Studies in Medical Ethics,* Harvard University Press, Cambridge, MA, (1977), pp. 17–21.

⁵Joseph and Julia Quinlan, with Phyllis Battelle (1977) *Karen Ann,* Doubleday, Garden City, NY, pp. 89–92.

⁶*See* Benjamin Freedman (1981) "One Philosopher's Experience on an Ethics Committee," *Hastings Center Report* 11 (April), 20–22; Joan Kalchbrenner, Margaret John Kelly, and Donald G. McCarthy (1983) "Ethics Committees and Ethicists in Catholic Hospitals," *Hospital Progress* 64 (Sept.), 47–51; and Robert M. Veatch (1986) "The Roles and Functions of Hospital Ethics Committees," *Ethics in the Hospital Setting: Proceedings of the West Virginia Conference on Hospital Ethics Committees*, Bruce D. Weinstein, ed., The West Virginia University Press, Morgantown, WV, pp. 1–15, especially p. 7.

⁷I develop this thesis in "DRGs and the Ethical Reallocation of Resources," *Hastings Center Report* 16 (June, 1986), 32–40.

⁸Robert M. Veatch (1976) *Value-Freedom in Science and Technology*, Scholars Press, Missoula, MT.

⁹*Elizabeth Bouvia v. Superior Court of the State of California for the County of Los Angeles* (Court Appeal of the State of California, Second Appellate District, Division Two, April 16, 1986).

¹⁰Joseph Fletcher (1966) *Situation Ethics: The New Morality* Westminster Press, Philadelphia, PA; cf. John Rawls (1971) *A Theory of Justice*, Harvard University Press, Cambridge, MA.

¹¹Mark Siegler (1979) "Clinical Ethics and Clinical Medicine," *Archives of Internal Medicine* 139, 914, 915 and "Cautionary Advice for Humanists" (1981), *Hastings Center Report* 11 (April), 19, 20.

The Skeptical Critique of Clinical Ethics

Robert Baker

PART ONE

Skepticism About Clinical Ethics

Socrates may have searched for knowledge in the *agora,* but our paradigm of philosophical activity is Plato speculating absolute truths in the relative isolation of Academe. As a consequence, pride of place in the intellectual world tends to be given to theorists, and they, in turn, have a tendency to be skeptical about those of us who, following Socrates, still seek knowledge in the marketplace, or to turn to the subject of this paper, in the clinic. In "Clinical Ethics, Applied Ethics, and Theory," and in an earlier essay, "The Ethics of Institutional Ethics Committees,"[1] Robert Veatch voices many of the skeptical concerns applied and theoretical ethicists have about the role of the ethicist in the clinic. That role, he argues, is inherently biased because the clinical ethi-

cist operates "on the clinician's turf, is paid by the health professional system, and gradually develops identification with
clinical professionals."[2]   More importantly, however, the
clinical ethicist's self-definition as advisor to clinicians is
logically incompatible with the basic bioethical ideal of self-
determination.

Veatch's Critique

The gist of Veatch's critique of the model underlying
clinical ethics is contained in the following passages:

> That model...lines the ethicist up on the side of the profes-
> sional and assumes that it is the professional, rather than the
> layperson, who is the decisive decision maker.
> Increasingly it is recognized that this is a serious mistake.
> Medical decisions necessarily involve ethical or other value
> components....This is true not only for stereotypically ethical
> issues....It is equally true for every routine choice that is made
> in the clinical setting....
> If this is so, then every clinical choice must be preferred on the
> basis of some value set. *Insofar as decisions are clinical, that
> is, ones that affect the layperson, the appropriate value set, at
> least to a large extent, is that of the layperson.* [PPD] There
> is no evidence whatsoever that professionals in a clinical
> field...have an expertise at choosing the values upon which
> clinical decisions must be based. Even if there were...*any per-
> son committed to the principle of autonomy would still prefer
> that, insofar as the decision affects only the patient, the
> patient's value system be used as the basis for the clinical
> choice.* [RPA]
> ...The implications of the above claims are far more radical
> than most people realize. If every clinical decision involves a
> value set and if the appropriate value set is not normally that
> of the clinician, then the ethicist in the clinical setting is
> making an error if he or she offers...advice... primarily for the
> clinician.[3]

Veatch's Appeal to Self-Determination

In the italicized sections of these quotations, Veatch appeals to two principles to generate his argument: the principle of patient determination and the principle of respect for patient autonomy. These principles can be formulated as follows:

PPD: The patient is the primary decision maker;
RPA: Clinicians have an obligation to respect patient autonomy.

Appeals to principles are characteristic of analytic moral theory. Like the name "applied ethics," the term "principle" suggests a body of ethical theory that can be applied, perhaps deductively, to resolve ethical difficulties in much the same way that engineers apply the principles of physics to resolve practical problems about building bridges and buildings. There is, of course, no such body of theory, but were there one, any North American variant would undoubtedly contain some version of PPD and RPA—if only because the rhetoric of self-determination appeals to the North American soul. More importantly, however, when elevated to the level of principle, the ideal of self-determination generates a minimalist ethic that readily accomodates the pluralistic nature of North American society. Pluralism implies an absence of authoritative moral principles and definitive decision procedures in matters of value. In such a context it appears eminently reasonable to allow individual moral agents to determine their own lives according to their own values and principles (except insofar as these impinge on others). Or, as Veatch puts this point in his defense of PPD, since "there is no evidence whatsoever that professionals in a clinical field...have any expertise in choosing the values upon which clinical decisions must be based," it follows that "the appropriate value set...is that of the lay person."

Veatch's invocation of PPD and RPA are part of the standard rhetoric of North American bioethics. The President's Commission for the Study of Ethical Problems in Medicine and Biomedical and Behavioral Research, for example, echoes similar sentiments in its 1982 and 1983 reports:

> ...decisions about the treatments that best promote a patient's health and well-being must be based on the particular patient's values and goals: no uniform objective determination can be adequate—whether defined by society or by health professionals.[4]

But unlike the President's Commission, which wholeheartedly endorses ethics committees[5] and the role of clinical ethicists, Veatch challenges conventional wisdom, arguing that the very principles of self-determination endorsed by the Commission and most bioethicists are incompatible with the idea of clinical ethicists. How does Veatch generate radical conclusions from such apparently innocuous and incontrovertible premises? Part of the answer lies in the presuppositions underlying his arguments; the other part, which I shall turn to next, involves Veatch's interpretation of autonomy and self-determination.

Autonomy, Self-Determination, Self-Control, and Self-Fulfillment

The honor of introducing the concept of autonomy into the lexicon of moral philosophy is generally accorded to the 18th century Prussian philosopher, Immanuel Kant.[6] Kant's conception of autonomy is essentially that of "self-control." Individuals are said to be in control of themselves, i.e., to act autonomously, if they govern their actions by their own ideals. Individuals are said to lose self-control, i.e., to act heteronomously, if they have no ideals for themselves, or if, having ideals, they lack the will to live up to them, or if some external agency prevents them from acting on their ideals. Thus, children, although they are free, are not autonomous (i.e., self-controlling) agents in Kant's view because they lack internal ideals with which to govern their actions. Agents with ideals, but without the will to live up to them, also lack autonomy. Neither dieters who succumb to temptation and break diets, nor, to take a more characteristically Kantian example, liars who fail to live up to their own standards of probity and break promises for self-serving reasons, are autonomous (self-controlling) agents in the Kantian paradigm. Both, acting on their desires (or inclinations), fail to live up to their own ideals, and, in succumbing to their inclinations, forfeit their autonomy. Some agents, of course, lose their autonomy to external pressures. They have ideals and the will to live up to them, but cannot do so because of external coercive forces. Any agent coerced to live or act according to another's ideals is also nonautonomous in a Kantian analysis.

When Veatch invokes the concepts of autonomy and self-determination (in his argument that in every clinician–patient encounter the dominant attitudes, interests, and desires ought to be those of the patient, not those of the clinician), he puts the issue in terms of the values governing choices, including routine medical choices. What he seems to have in mind here is the attitudes, interests, and desires of clinician and patient. He does not talk of ideals or objectives, so the operative notion of autonomy appears to be subtly different from the Kantian notion of autonomy as self-control. It is perhaps best understood as self-fulfillment (i.e., the fulfillment of personal desires, and the satisfaction of individual interests and attitudes.)

The difference between the self-fulfillment and self-control interpretations of autonomy are not significant in cases involving coercion or domination; slaves, for example, are paradigms of heteronomy in either interpretation of autonomy or self-determination. Both interpretations of autonomy are profoundly antipaternalistic, and both endorse the primacy of the patient's right to refuse treatment. The difference between the two interpretations emerges when we consider children, weak-willed dieters, and self-serving promise-breakers. For Kant, these are paradigms of heteronomy because, even though they involve choices made on the basis of desires or interests, they represent surrenders of autonomy, not assertions of autonomy. In the self-fulfillment conception of autonomy, however, it is inconceivable (if not self-contradictory) that a choice based on the agent's own desire could be nonautonomous. Any choice made on the basis of the agent's own values (desires, wishes) must be regarded as an instance of autonomous agency (including the decision to break a diet or otherwise abandon one's ideals).

The self-fulfillment interpretation of autonomy leads, by a fairly direct argument, to the rejection of ethicists and ethics committees, for, in the self-fulfillment interpretation of autonomy, anything that frustrates the realization of a patient's desires or wishes infringes the patient's autonomy. Thus, insofar as ethics committees and clinical ethicists may reinforce clinical recalcitrance, they infringe patient autonomy, and so, as Veatch argues, their role is incompatible with a commitment to patient self-fulfillment.

In the remainder of Part One, I criticize another assumption that Veatch makes. In Parts Two and Three, I will challenge the self-fulfillment interpretation of autonomy. My point is not that a self-control interpretation of autonomy is fundamentally correct, but that certain features of this interpretation—specifically the idea that values are complex, structured, and hierarchical—more aptly describe decision making in the medical context than does the self-fulfillment interpretation. In Part Two, I shall argue that the self-fulfillment interpretation of autonomy is unreasonable in an in-patient context. In Part Three, I will attempt to show that ethics committees and clinical ethicists are compatible with a more complex interpretation of autonomy and that, in fact, they typically act to enhance patient autonomy.

The Adversarial–Asymmetrical Model of the Clinical Encounter

To generate his critique of clinical ethicists and ethics committees, Veatch presupposes more than a self-fulfillment conception of autonomy. He also presumes an adversarial model of the clinician–patient encounter. Thus, he claims not only that "every clinical choice must be preferred on the basis of some value set," but that the value sets of clinician and patient differ significantly, that both the clinician and the patient seek to shape clinical decisions to conform to their own values, and that each party vies with the other to control the outcome of the encounter.

Adversarial models of the clinician–patient encounter are commonplace in bioethical theory, perhaps because bioethics arose largely in response to a number of famous legal cases (*In re Quinlan, Kaimowitz v. Department of Mental Health, Roe v. Wade, Superintendent of Belchertown v. Saikewicz, Tarasoff v. University of California*, and so on), and court cases are inherently adversarial. Reading the bioethical literature sometimes leaves one with the impression that all clinician–patient encounters are variations on adversarial themes—it is almost as if every outpatient consult involved a Jehovah's Witness and a surgeon.[7]

Adversarial encounters typically generate "winners" and "losers." Thus, this model of the encounter naturally prompts Veatch, the President's Commission, and other bioethicists who presuppose it to query, "Whose values should predominate in clinical encounters?" Veatch replies (as does the President's Commission) that the values of the party or parties most affected by the decision ought to shape the decision (PAD). Given a (typically unstated) presupposition that patients are more significantly affected by clinical decisions than clinicians, Veatch (and the President's Commission) are led to the further conclusion that "the appropriate value set...is that of the layperson."

Is the Clinical Encounter Inherently Adversarial?

The adversarial model of the clinician–patient encounter, however, is profoundly misleading. Andy Rooney, the American Everyman, agrees with Veatch that "people don't like to go to see their doctors." But, he suggests, it is not because the situation is inherently adversarial; to use Rooney's words, it is "not because people are afraid of [doctors]." Rather, people avoid doctors because "they know they are going to have to take their clothes off."[8] Rooney's remarks are funny because they are true, and because they skirt a larger truth—that we are afraid to go to doctors because we fear what they might tell us about ourselves.

Consider, however, Rooney's lesser truth—our dislike of undressing. What is at issue is not necessarily modesty (for we strip in the locker room) but what Veatch might, perhaps, characterize as our surrender to the alien value set of the clinic. This value set is aptly represented by those formless, backless, anonymous garments known as hospital gowns. Such gowns were designed to serve the purpose of the clinic (to facilitate physical examinations, the use of bed pans, and so on). It is difficult to imagine anyone liking patient gowns or choosing to wear them of their own accord, for, not only do they affront our modesty and dignity, they also homogenize us into relatively anonymous objects of clinical procedure. Thus, wearing a hospital gown would appear to be a minor instance of the sort of normatively adversarial encounter Veatch has in mind—the gown embeds values that are at odds with those we normally act on, and patients wear them only under coercive pressure from clinicians.

Actually, as we all know, there is another interpretation of why patients wear hospital gowns, an interpretation that presupposes a more cooperative model of the clinical encounter. We don hospital gowns, neither because we change our sartorial values when we enter the doctor's office, nor because we submit to the alien values of the clinic. In the clinic we neither dress for success, nor to impress others with our status, taste, or uniqueness; instead, we dress to expedite a physical examination, and we submit to the examination because we are concerned about our health (and we expect the nurse, physician, or surgeon to share this concern). Insofar as everyone acts toward this goal (patient health), there is no normative conflict and the encounter is not adversarial.

The appearance of an adversarial encounter tends to be sustained by the self-fulfillment model of autonomy. This model allows little room for the complex, prioritized structure of ideals, goals, interests, desires, wishes, and whims that governs most people's lives. Everything is homogenized under the simple rubric of "values," hence any overridden value lends a false appearance of normative conflict to clinician–patient encounters. Thus, our dislike of hospital gowns on sartorial grounds suggests a conflict of values when it is merely evidence, not of an adversarial encounter, but of value prioritization.

Clinical encounters become truly adversarial when objectives conflict *and* when, in addition, one party seeks to impose his or her objectives on the other. In the Jehovah's Witness case, for example, the encounter becomes adversarial only if the surgeon is primarily concerned with the patient's physical malady and the patient is more concerned with a relationship with God than with physical survival. (The Jehovah's Witness interpretation of *Genesis* **9**: 3–4, *Leviticus* **17**: 13–14, and *Acts* **15**: 19–21—"Abstain from ...fornication and from what is strangled and from blood"— requires Witnesses to abstain from blood transfusions). This conflict of priorities becomes adversarial only if, in addition, one party attempts to impose his or her values on the other. Here the encounter becomes a bioethical textbook case, a zero-sum game, a prize fight in the arena of values. The party who achieves the higher priority in the encounter can be said to "win," whereas the other party, the adversary, "loses."

Standard clinical encounters, however, are not adversarial in this way. Both clinician and patient are generally thought to have "won" (if that term is even appropriate to describe the outcome of a clinician–patient encounter) when the patient can be pronounced "healthy." Wearing a hospital gown is neither a victory for the clinician nor a loss for the patient; it is merely an expeditious aid to diagnosis and treatment (to provide access for a stethoscope or a bed pan), an awkward expedient to a mutual goal. As I shall argue in Part Three, even the Jehovah's Witness case need not be (indeed should not be) conceptualized as an inherently adversarial encounter because to impose an adversarial conception on the encounter distorts its nature and prevents a resolution of the problem.

The Presumption of Asymmetry

The adversarial paradigm of the typical clinician–patient encounter is, I suspect, capable of exerting persuasive power only if it is unarticulated. Most philosophers, however, would be inclined to endorse another element of Veatch's model of the clinician–patient encounter—its asymmetrical impact. They would tend to agree that most clinical encounters, including physical examinations, impinge more significantly on patients than on clinicians. Physical examinations, however, are outpatient procedures. Clinical ethicists work, not in the outpatient world, but in the universe of 300–900-bed teaching hospitals, where most patients with whom they deal are inpatients. This world is radically different than the familiar world of the doctor's office. It is a world of *clinicians* rather than physicians, a world of medical technicians, respiratory therapists, nurses (practical nurses, licensed public nurses, and registered nurses), physician assistants, medical students, interns, residents, fellows, attending physicians, and head physicians, as well as private practitioners (including the patient's doctor if the patient has a private doctor). It is, of course, unreasonable to presume that paradigms developed on the basis of outpatient encounters apply in this world.

One of the strange features of the inpatient universe is that the asymmetry of effect sometimes reverses itself, i.e., clinicians are sometimes *more* significantly affected by clinical decisions than patients. The medical sociologist Charles Bosk describes this phenomenon in his memorable study of surgical units, *Forgive and Remember*.[9] Clinicians, he reports, are held strictly accountable for routine clinical decisions (deciding the frequency and dosage of medications, ordering preliminary and precautionary tests, and so on). They are expected to decide these matters in ways that optimize the odds in favor of their patients' recoveries. Any decision or action that appears nonoptimal is subject to public challenge, usually at staff rounds. These challenges have an enormous effect upon a clinician's standing and reputation. A staff member who is unable to justify a perceived technical transgression (e.g., a failure to change bandages) will be required to admit the mistake openly and not to repeat it; transgressors who fail to pronounce their *mea culpa* publicly, who persist in their mistaken ways, or who commit moral rather than technical transgressions (e.g., placing personal interests above patient interests) will not be forgiven.

Clinicians who commit unforgiveable mistakes often suffer extreme penalties, even in cases in which their patients suffered no harm. I remember a case in which an intern failed to perform a routine, entirely precautionary procedure. She cashed a paycheck instead of taking a patient to radiology for precautionary CAT scans. Although the chance of such a scan revealing brain damage was extremely slight (and when the CAT scan was performed two days later, it came out with the expected negative results), the intern was considered to have abandoned her patient. The attending physician upbraided her in front of the entire unit ("Do you intend to make a habit of ignoring your patient's needs, Doctor?"), and the intern was ostracized by the attending physician and by the house staff for the remaining weeks of her rotation through that unit.

It may seem strange to the casual observer that an intern be so severely penalized for such an apparently minor transgression, but clinicians use routine decisions as a mechanism for assessing the character of the decision maker. (Or, if one prefers to interpret their actions in consequentialist terms, they use it to assess patterns of decision making). Thus, when they chastise a fellow clinician, they are addressing a failure of character (or attempting to revise a pattern of decision-making behavior), and relatively dire remedies appear to be justified even when the transgressions had only a marginal impact on particular patients. Consequently, although the unchanged bandage or the delayed CAT scan might affect a patient significantly, it is much more likely to lead to the chastisement of the clinician. In the inpatient universe of the clinic, therefore, insofar as the effects of routine decisions are asymmetrical, they are more likely to affect clinicians than patients.

PART TWO

In this part of the paper, Veatch's critique of the role of clinical ethicists will be assessed in light of the analyses of Part One. My procedure will be to use the theoretical perspective of a tripartite Rawlsean contract between clinicians, patients, and a society that Veatch made famous in *A Theory of Medical Ethics*.[10] (Unlike many applied ethicists, Veatch has been careful to eschew the practice of principle prestidigitation—in my opinion, the bane of analytic bioethics. He usually tests his arguments by developing them within the theoretical framework of a hypothetical social compact between clinician, patient, and society modeled on the contracts John Rawls analyzes in *A Theory of Justice*[11]). In this part I submit my appraisal of his argument to the same discipline.

A Contractarian Reassessment
of Veatch's Critique

Recast within a contractarian framework, Veatch's critique of clinical ethicists boils down to two claims: (1) Any societal–patient–clinician compact must embrace a principle (either PPD or a version of RPA) entailing that clinicians are obligated to execute all patient requests except those "so morally repugnant that the clinician cannot in good conscience continue; provided, within certain limits, that other professionals are available to step in and provide care"[12]; and (2) If the compact contains either of these clauses, clinicians are secondary parties in clinical decision making; hence, ethicists are at best ancillary to the secondary, which is to say, irrelevant.

Veatch appears to assume that all parties to the hypothetical compact would agree on either PPD or his interpretation of RPA. Here again, however, one's presuppositions about the clinician–patient encounter are the key to the argument, for although patients who conceptualize these encounters on the model of outpatient consults (which is, after all, the primary experience most people have with clinicians) might be tempted to accept a self-fulfillment interpretation of these principles, that temptation would vanish if their model of the encounter embraced the hospital ward. Hospital wards are shared public facilities. The patients who occupy these facilities at a given time often have conflicting values, interests, and desires. If their rights to self-fulfillment were given free rein, these conflicting interests would express themselves as a classical Hobbesean war of each against all, a situation ultimately undesirable for everyone.

Worse still, Veatch's interpretation of PPD would be utterly unworkable for inpatient clinicians, e.g., a nurse. A nurse responsible for a typical semiprivate room with two to four patients, central lighting, a radio, a television, and only one heating control, presides over a domain with the potential for innumerable disputes (about lighting, heating, music, television, being too bright/dim, too hot/cold, too loud/low, or too quiet/disturbing). The impact of Veatchean patient-determination (PPD) would appear to require the nurse to act as a mere executor of patient requests. Since these requests are likely to be incompatible, the nurse would be immobilized and Hobbesean chaos would triumph. Rational patients would, of course, wish to resolve this situation by appointing the nurse as either a Hobbesean sovereign or a Lockean umpire; that is, they would gladly relinquish their unfettered rights of self-fulfillment (and embrace an alternative interpretation of PPD and RPA).

Thoughtful patients (and a liberal society) might also be reluctant to accept Veatch's rationale for PPD and for his interpretation of RPA because he supports his self-fulfillment interpretation with the argument that, insofar as a party is affected by a decision, that party's values ought to determine this decision [PAD]. As we observed in Part One, however, on inpatient wards routine decision making impinges asymmetrically, typically affecting *clinicians* more significantly than patients (because of the stringent standards of accountability to which clinicians adhere). Thus, if one accepts the validity of PAD, it seems to follow that, at least in routine clinical decisions, the predominant value set ought to be that of *clinicians*, not patients. Yet I doubt whether patients would willingly compromise their rights of privacy, i.e., their Kantian rights of self control (their rights to refuse interventions), on the basis of PAD.

PAD is actually a self-control compromising principle. Organ recipients, for example, are more profoundly affected by the decisions of potential organ donors than the potential donors themselves (at least in cases where the organ in question is redundant and where failure to receive an organ is fatal). Thus, applying Veatch's line of reasoning to this situation, it would seem that it is the values of the potential recipients, *not the donors*, that should determine whether a potential donor ought to donate an organ. It would also appear that potential recipients who accept Veatch's argument (and whose value systems would have committed them to giving organs to others) would have the right to coerce potential donors into becoming actual donors (unless, of course, the recipient's value system disallowed coercion). Thus, PAD could be used to override the autonomy not only of clinicians and organ donors but, depending upon the context, of any member of society.

A liberal society will not accept PAD. The fundamental problem with PAD is that the notion of self-determination is interpreted as giving individuals positive rights against others in order to achieve self-fulfillment, at least when their interests are stronger than those of others. This establishes an affective or interest utilitarianism that overrides strong negative rights to privacy, i.e., to self-control. Classical liberal theory is more traditionally Kantian, emphasizing negative, privacy rights that effectively block the intrusion of the interests of other parties. By embracing an argument favoring the self-fulfillment or the interest rights of patients over the privacy or the self-control rights of clinicians, Veatch effectively embraces an anti-liberal principle.

Clinicians, as the targets of the argument, would never accept Veatch's interpretation of PPD or RPA, at least not as a clause in a voluntary tripartite compact. No labor union would accept similar conditions for its workers. If, for example, we were talking of "passengers" rather than "patients" and "airline attendants" rather than "medical attendants," no self-respecting attendant's union would sign a contract in which attendants must comply with passenger requests except when (to quote Veatch) they are "so morally repugnant that the [attendant] cannot in good conscience continue; provided, with certain limits, that other professionals are available to step in and provide [the service]." Not only could the attendants not run an airplane under these conditions (given the potential for conflicts between passengers), they would never consent to conditions of petty passenger tyranny. Similarly, no clinician's union would accept petty patient tyranny.

As Veatch remarked in *A Theory of Medical Ethics*:

> Those who have reflected on this problem have concluded that individual autonomy is very basic. It is to be protected even at some loss in terms of other, nonmoral goods. John Rawls, whose approach, like the one being used here, is contractarian, suggests that the principle of liberty would be the first principle of a contractarian system. As he puts it, rational contractors from behind a veil of ignorance would agree that "each person is to have an equal right to the most extensive basic liberty compatible with a similar liberty for others."[13]

The fundamental principles governing clinician–patient encounters in a tripartite Rawlsean social compact need to recognize and reconcile claims to autonomy on the part of *all* parties to the compact. PPD and Veatch's interpretation of RPA fail to meet this standard, for each fails to recognize the autonomy of clinicians and their need to act as accountable moral agents. It is not my purpose here to outline a more adequate set of principles. I do believe, however, that a non-adversarial model of the clinician– patient encounter is the proper framework for the development of precepts that will accommodate the self-respect and autonomy of both clinician and patient.

PART THREE

Veatch's critique fails to provide sufficient grounds for excluding the role of clinical ethicist from a clinician–patient–societal compact. The role of clinical ethicist is thus morally possible. But is it also morally desirable? Would any of the parties to the compact have a good reason to request, or even to insist on, a role for clinical ethicists? Would any of the signatories veto such a role? Would patients object to the idea of ethicists because, as Veatch remarks, the ethicist "is on the clinician's turf, is paid by the health professional system, and gradually develops identification with clinical professionals?" To address these questions, one must first assess the roles that clinical ethicists currently play and the context in which they might play them.

The Roles Clinical Ethicists Play

Clinical ethicists generally function in a medical-center teaching-hospital context in acute care settings. It is a community of hierarchically structured health care teams who administer their responsibilities collaboratively, holding others accountable for their individual actions and the actions of their subordinates. It is a community in which patients, though residing in private or semiprivate rooms, share common spaces, common personnel, common services, common equipment—and common problems. It is a community that, although dedicated to the autonomy of patients over their own bodies and to honoring the autonomy of clinicians in the workplace, is continually dealing with compromised autonomy, because the essentially heteronomous processes of disease naturally compromise the autonomy of patients, whereas the bureaucratic procedures essential to the mass production of services constrain the autonomy of both clinicians and patients. Patients, however, typically suffer a deeper erosion of autonomy, in part because of their diseases and disabilities, but in large measure because the bureaucratic apparatus of the clinic tends to strip them (but not clinicians) of the accoutrements integral to asserting their autonomy. (The anonymous hospital gown is a case in point). The world in which the clinical ethicist typically functions, then, is not the sanctum sanctorum of the doctor's office; it is not an isolated world in which clinician and patient face each other alone as autonomous equals. It is a quintessentially communal world, where clinicians act collaboratively and semi-autonomously; it is a world in which patients are fortunate if they can assert any autonomy at all. It is a world in which clinicians stand as semi-autonomous verticals, whereas patients lie supine as heteronomy-engulfed horizontals.

Is there a role for the clinical ethicist in this world? Ethicists have, in fact, played eight roles in this environment. Seven of these roles can be characterized (following Veatch's alliteration) with a word starting with the letter 'A': activator, articulator, analyst, advocate, æducator, advisor, and arbitrator. The eighth function is *educator*. Most medical centers train clinical professionals; one of the primary functions almost all clinical ethicists serve is to formally instruct trainees (medical students, nursing students, and students of allied health care professions) about the ethical aspects of the professions they are about to enter. Ethicists continue to serve as an educational resource for clinicians by offering postgraduate sessions (lectures, minicourses) on current issues of moral concern. As teachers concerned with moral questions, clinical ethicists become exemplars of the idea of morality in the minds of clinicians. The association allows them to play two key roles within the clinic, that of arbitrator and activator.

Like most complex institutions, clinics are prone to intra–staff disputes. Ethicists, because of their relative neutrality in the territorial and hierarchical arrangements of clinical institutions, and because of their identification with the idea of ethics, are often appealed to (as neutrals) to arbitrate any number of disputes, many of which are only remotely ethical in nature.

Although arbitration is not a major role for most clinical ethicists, it is worth mentioning because it underlines the close association between the idea of ethics and the person of the clinical ethicist, at least in the mind of clinicians. This identification permits ethicists to play a more subtle and important role in the clinical world, the role of activator of moral concern. The clinical staff at a medical center are in the business of mass-producing complex services, so they naturally tend to be preoccupied with the pragmatics of routine. This preoccupation often leaves them little time for reflection on medicine, much less on the morality of medical practices. To counter this preoccupation with the routine, most clinical facilities schedule reflective enclaves known as "rounds" into their working day (morning rounds, afternoon rounds, patient rounds, staff rounds, nursing rounds, attending rounds, radiology rounds, pathology rounds, grand rounds, and so on). During the brief, reflective enclaves created by rounds, options are assessed, complex decisions are reviewed, and the hierarchical structure of the clinic is partially suspended to permit relatively open discussions (in which subordinates have the temporary prerogative of questioning, and even politely challenging, their nominal superordinates).

Clinical ethicists play three activator roles within the context of rounds. First, by their very presence in the clinic, they remind clinicians that moral considerations are as integral to the clinical practice of medicine as caring and curing.

Secondly, clinical ethicists, by their very presence as visiting "experts" at regular unit rounds, activate the discussion of ethical issues at these rounds. (The dynamics of the activation function in the context of rounds is interesting. Etiquette requires rounds participants to focus either on the issues most germane to the topic of the rounds or the specialties of visiting clinicians. When an infectious disease expert "rounds" with a unit, discussion tends naturally to focus on questions of infection; similarly, when a clinical ethicist "rounds" with a unit, since the ethicist's expertise is "ethics," discussion naturally turns to the ethical aspects of the cases under discussion). Finally, ethicists sometimes organize "ethics rounds," i.e., rounds that focus entirely on the moral aspects of medical practice.

Clinical ethicists not only activate discussions of the moral dimensions of medicine within the clinic, they assist in the articulation of the positions of clinicians. Stanley Reiser, a medical historian and clinical ethicist at the University of Texas Medical Center in Houston, is my personal paradigm of the ethicist as articulator. His technique is beguilingly simple. Using the format of clinical rounds, he encourages physicians to review troublesome cases retrospectively. After listening to the case presentation, he patiently uses silence, a few questions, chalk, and a blackboard to elicit from clinicians the nature of *their* concerns about the case—allowing them to discover what *they* believe to be the crucial points at issue. The articulation process enables clinicians to appreciate the specific elements that made a case problematic, thus allowing them to consider how best to alter their procedures to prevent similar problems from arising in the future.

Articulation is prolog to analysis in the minds of philosophical theorists (but not necessarily in the more practically oriented mind-set of the clinician, who is often content to revise practices in a manner that will prevent the recurrence of problematic cases). Since ethics rounds are often led by clinical ethicists with extensive training in moral philosophy and/or moral theology, analysis frequently follows articulation. But these analyses tend to be subtly different from those of bioethical theorists, for, as Veatch observes, clinical ethicists "gradually develop identification with clinical professionals." Thus, when clinical ethicists conduct analyses at rounds, they tend to eschew the notoriously recondite rhetoric of applied ethics ("deontological,""nonmaleficence," "supererogatory," and so on) and express themselves in a language more accessible to the clinical perspective. Perhaps more importantly, clinical ethicists bring a clinical perspective to the applied ethics literature, a perspective that suggests models, paradigms, and presuppositions that might be overlooked from the armchair perspective of the applied ethicist or the ethical theorist.

Clinical ethicists can do more than activate moral concern, articulate, and analyze; sometimes they use their position as a "bully pulpit" to acquaint clinicians with aspects of the applied ethics literature that they believe to be particularly relevant to a clinical unit's actual practices. In keeping with the spirit of Veatch's alliteration, I shall designate this extremely tendentious insertion of moral theory into clinical practice active-education, or *æducation.*

My personal paradigm of the clinical ethicist as æducator occurred several years ago when I was doing research on an intensive care unit at a large urban medical center. During my stay, I had occasion to observe the center's clinical ethicist at ICU case rounds. At his request, the clinicians presented the case of Mr. Old. After the presentation, the ethicist—who was experienced enough to evoke the non-accusatory reflective ethos of clinical rounds—"presented" some literature on discrimination and asked the staff to consider whether, according to the criteria in the published applied ethics literature, their DNR order for Mr. Old was an instance of discrimination (ageism). What followed was a vigorous but unemotional discussion in which the staff—which readily admitted using age as a factor in their decisions—tried to assess the justifiability of their treatment criteria in light of the literature on discrimination. It was a remarkable event. The ethicist was, in effect, accusing the staff of allowing the death of a patient on discriminatory grounds; the staff was dispassionately considering the accusation. This type of dispassionate reflection is possible only because of the institution of clinical rounds, and because of the uniquely æducational role the ethicist can play within that context. It is noteworthy that at the end of rounds, while the ethicist was handing out some offprints and a bibliography, the unit head and several of the staff thanked him for presenting rounds—and reminded him that he was scheduled to present again in three weeks.

The clinical ethicist's sphere of influence extends from the wards and the lecture hall to meeting rooms. In meeting rooms they serve as advisors on the policymaking committees of medical centers (admissions committees, animal subjects research committees, continuing education committees, curriculum committees, ethics committees, graduation committees, grievance committees, human subjects research committees, neonatal ethics committees, and so on). In this capacity they can shape the clinical experiences of patients. Consider, for example, a policy matter recently discussed by the ethics committee of a nursing home. The home found itself routinely transferring chronically ill patients to the emergency rooms and intensive care units of large medical centers for the last few days of their lives. The staff believed that these patients would be better off dying in their own rooms, visited by their friends, and cared for by a

staff who knows them and their families. Yet in the absence
of specific directives from the families or the patients, they
felt no choice but to transfer these patients whenever they
became acutely ill.

When they presented their views to the ethics committee, I suggested (as the institution's clinical ethicist) that
they develop a prospective Do Not Transfer (DNT) policy.
DNT protocols are similar to standard DNR protocols, except that the subject of a DNT order is a decision not to transfer a patient to an acute care center instead of a decision not
to resuscitate. (In fact, however, a decision not to transfer a
patient from a nursing home to an acute care facility is
usually tantamount to a decision to forgo resuscitative technology.) Approximately six months later the policy was
implemented; a discussion of DNT has been incorporated
into the three-week orientation program the home runs for
all newly admitted patients and their families; and all
patients and families are asked to consider the possible circumstances in which it would be appropriate to transfer the
patient to an acute care facility. The program is only a few
months old, but it seems to be working to everyone's satisfaction. The point I wish to underline here is that advising in
an institutional context is not a matter of taking sides in an
adversarial encounter between clinician and patient; it is
more typically a matter of restructuring a clinical situation
in a manner responsive to *both* clinician concerns and
patient interests. In the case in point, we were able to widen
the sphere of patients' reflective choices, thereby expanding
their ability to control their lives and the manner of their
deaths.

Sometimes, more often than clinicians would like to
admit, clinical processes are directly inimical to patient welfare. In such circumstances, clinical ethicists often play the
role of advocates *for morality*. Note that this is not, in any

direct sense, a patient–advocate role; rather, it is the role of advocate for the moral point of view. For example, every Tuesday the orderlies at a teaching hospital in a large urban medical center would array all the women waiting second trimester abortions in a neat line in front of the delivery room (where the abortions would be performed). Unfortunately, the corridor was flanked on either side by glass-walled nurseries so that women awaiting abortions were left to stare, often for hours, at newborn babies. The clinical ethicist working at the medical center noticed the situation (actually he noticed a distraught, sobbing woman trying to turn away from the babies). He discussed the woman's distress at obstetrics/gynecology rounds later that month, and the clinical staff decided to eliminate the queuing of abortion patients in the public corridor. Several months later, he asked the obstetrics/gynecology staff how the new procedure was working out. There was universal agreement on the superiority of the new arrangement. One nurse remarked, "I can't think why we didn't always do it this way." Although her grammar is awkward, her meaning is straightforward. Clinician–patient interactions are not inherently adversarial; so ethicists, acting as advocates for morality, need not side with either clinician or patient. More often than not, the morally desirable solution is optimal for all parties.

A Contractarian Defense of the Role of Clinical Ethicist

Consider again the question of whether any party to the Veatchean tripartite compact would request the presence of ethicists in the clinic. Clinician–patient encounters in the hospital context are morally complex. Even in a morally ideal hospital—a hospital in which, among other things, all

clinicians genuinely respected the autonomy of all patients—the clinician–patient encounter would still generate
moral controversies. Disease, as a process, is not respectful
of patient autonomy; it is essentially a state of heteronomy,
i.e., a dominance of body over will. The essential heteronomy
of disease inevitably provides plausible grounds for clinical
paternalism. Clinicians can properly reason that patients
committed to their own autonomy (i.e., to their health)
should act in a manner that optimizes that priority. (To cite
Kant's dictum, "To will the end is to will the means.") Thus,
someone seeking longevity should cease smoking cigarets,
diabetics with similar aims should stick to their diets, and,
for the same reason, hospitalized patients should follow
their clinicians' recommendations. To do otherwise is to act
inconsistently with the will to health and, rather than dis-
playing autonomy, merely evinces heteronomy. Even the
most moral clinician operating in a morally ideal hospital
will always have plausible paternalistic reasons to consider
overriding, ignoring, or manipulating the behavior or wishes
of any patient.

Patients, of course, have morally and logically sound
reasons to reject such manipulations. They can always
claim that other considerations override their concern for
health (e.g., the Jehovah's Witness's spiritual concerns).
Thus, the clinician–patient encounter, although it is not
inherently adversarial (since clinicians and patients share
the same objective—the health of the patient), is inevitably
subject to different interpretations. Patients are naturally
inclined to interpret their hesitations about clinical recommendations and procedures as expressions of their autonomy (rather than failures of nerve or weakness of will);
clinicians, on the other hand, are occupationally predisposed
to regard them as expressions of patient heteronomy and
ignorance (rather than considered reflections of patient au-

tonomy). The complexity of the interpretation and the natural biases of the parties involved create a need for clinical ethicists.

When clinical ethicists play their various roles as educators, arbitrators, activators of moral concern, articulators, analysts, æducators, advocates of morality, and advisors, they create opportunities for clinicians to reflect on their natural interpretations of their activities vis-à-vis patients. Thus, it might appear that, as Veatch notes, clinicians are the beneficiaries of ethicists' advice. But consider: who benefits from the ethicists' intervention? Is it the semi-autonomous, vertical clinician, or the supine, horizontal, bureaucratically-controlled, heteronomous patient? When the interpretations of clinician and patient differ, whose interpretation is likely to prevail? Who, then, has the most to gain by having an external, relatively neutral party review clinician and patient interpretations of a situation? The patient would seem to have little to lose and everything to gain from the presence of a neutral party who can activate moral concern on the part of the more powerful, vertical clinician. Even the ethicist who played the role scripted by Veatch, i.e., even the ethicist who functioned merely as a redundant echo of clinical values, would still serve the patient's interests well. For merely by being a presence in the clinical complex, the ethicist inspires a concern for, and a discussion of, moral issues that are often ignored—typically to the patient's detriment.

The clinical ethicists with whom I am acquainted, however, do more than inspire open discussions of morally sensitive issues. They can, and do, challenge clinicians to reconsider their interpretations of cases (as the ethicist did in the case of Mr. Old); they can, and do, challenge clinicians to reconsider the moral implications of their practices (as in the case of queuing abortion patients in front of nurseries); and they can, and do, advise clinicians on how to restructure the clinic to be more responsive to patients' assertions of autonomy (as in the case of the DNT protocol). Patients benefit directly from these endeavors. Therefore, insofar as patients act in their own self-interest, they have every reason to insist on clinical ethicists as part of the tripartite compact.

Society's interests coincide with the patient's in this matter. Society has every reason to seek morally optimal outcomes and hence to support the role of the clinical ethicist.

Clinicians, then, are the only parties to the compact who might have reason to resist the clinical ethicist. If they viewed the clinician–patient encounter as inherently adversarial, they would resist. They do not, and, as I have argued above, should not regard the encounter as adversarial. Most clinicians are sincerely committed to restoring autonomy to their patients and, because they also believe that ethicists can assist them in attaining this end, ethicists have been welcomed into clinics all over North America.

The Case of the Jehovah's Witness Reconsidered

Consider again the textbook case of the Jehovah's Witness and the surgeon. What is intriguing about this encounter is not the potential conflict of values, but the fact of the clinical consult, i.e., the fact that the Jehovah's Witness visited the surgeon at all. The visit suggests a temporal concern with bodily affairs that can be weighed against spiritual priorities. A case similar to this was raised at recent surgery rounds. The surgeon's initial attitude was that patients had no right to force him to operate "with one hand tied behind my back." The hospital's ethicist presented an alternative view of the situation at rounds the next day by leading a discussion based on an excellent article on Jehovah's Witnesses by Dixen and Smalley,[14] which explains Witness theology and analyses the medical alternatives open to surgeons (including interoperative salvage and artificial blood). The ethicist also reminded the surgeon of the obligation to help patients. Soon the session focused on the technical challenge involved in avoiding transfusions.

In the event, the actual surgery proved anticlimactic. The patient complicated matters less than most Witnesses because she did not object to a transfusion of her own banked blood, despite the fact that autotransfusions of predeposited blood are contrary to Jehovah's Witness doctrine. Her surgery proved to be medically, theologically, and morally uneventful. Although her own blood had been banked, and although recapture devices were deployed throughout her surgery, the precautions the surgeons took to minimize bleeding were so effective that there was no need for a transfusion.

Jehovah's Witnesses cases, which seem intractable if one's conception of the clinical encounter is filtered through adversarial, self-fulfillment paradigms, can be resolved by exploring the complexities of value structures and the potential for cooperation they allow. A compromise can usually be reached that does not compromise the fundamental values of either clinician or patient. The role of the clinical ethicist and the ethics committee is first, to create the reflective environment in which such compromises can be sought, and secondly, to work within that environment to achieve them. Every party to a Veatchean tripartite compact should welcome the solution achieved in the case of the Jehovah's Witness. Therefore, every party to a Veatchean compact has reason to accept the clinical ethicist and the ethics committee—which suggests that, even when measured against the stringent standards set by Veatch, ethics committees and clinical ethicists are more than morally justifiable, they are morally desirable.

Acknowledgments

I would like to thank the editors for their comments on a draft of this essay and to express my gratitude to my friend and colleague, Stanley Kaminsky, for reading the various drafts of this essay and for his invaluable suggestions and criticisms.

References

[1]Robert Veatch, "Clinical Ethics, Applied Ethics, and Theory," *infra* this volume, chapter 1, pp. 7–25 and "The Ethics of Institutional Ethics Committees," *Institutional Ethics Committees and Health Care Decision-Making* (1984), Cranford R. and Doudera A., eds., (Health Administration Press, Ann Arbor, MI), pp. 35–49.

[2]R. Veatch, "Clinical Ethics, Applied Ethics, and Theory," *supra*, note 1, p. 20.

[3]*Ibid.*, pp. 17,18 (italics added).

[4]President's Commission for the Study of Ethical Problems in Medicine and Biomedical and Behavioral Research (1983), *Deciding to Forego Life-Sustaining Treatment* (US Government Printing Office, Washington, DC), p. 43. *See* also the Commission's report, *Making Health Care Decisions* (1982) (US Government Printing Office, Washington, DC).

[5]A number of different types of committees are referred to as "ethics committees": advisory–consultative committees (AHA ethics committees), allocation committees, immunity-conferring committees (Teel-type committees), watchdog committees (e.g., ICRCs, IRBs), and so on. Throughout this paper I use the term to designate only advisory– consultative committees of the type recommended by the American Hospital Association in its "Guidelines on Hospital Committees in Biomedical Ethics" (reprinted in Cranford and Doudera, *supra*, note 1, pp. 403,404).

[6]Immanuel Kant, *Groundwork of the Metaphysics of Morals*, translated by H. J. Paton (1964) (Harper and Row, New York, NY).

[7]Cf., e.g., Case 105 in Robert Veatch (1977) *Case Studies in Medical Ethics* (Harvard University Press, Cambridge, MA), pp. 327–332. It is worth noting that in Veatch's *A Theory of Medical Ethics* (1981) (Basic Books, New York, NY), his arguments for his theory center on an analysis of twelve cases, almost all of which are given an adversarial interpretation.

[8]A. Rooney (1986) *Word for Word* (Putnam, New York, NY), p. 19.

[9]Charles Bosk (1979) *Forgive and Remember* (University of Chicago Press, Chicago, IL).

[10]R. Veatch, *A Theory of Medical Ethics, supra,* note 7.

[11]John Rawls (1971) *A Theory of Justice* (Harvard University Press,Cambridge, MA).

[12]R. Veatch, "Clinical Ethics, Applied Ethics, and Theory," *supra,* note 1, p. 24. *See also* Veatch, *A Theory of Medical Ethics, supra,* note 7, p. 198.

[13]R. Veatch, *A Theory of Medical Ethics, supra,* note 7, p. 194.

[14]J. L. Dixen and M. Smalley (1981) "Jehovah's Witnesses: The Surgical Ethical Challenge," *JAMA* **246,** 2471,2472.

Moral Experts
and Moral Expertise

Do Either Exist?

Arthur L. Caplan

The Growth in the Utilization
of Moral Experts

Recently a brochure was sent to a variety of health care
professionals and others interested in medical ethics an-
nouncing the existence of a new newsletter. "One bad ethical decision could destroy your career," the recipients were
warned. In order to avoid this dire fate, prospective cus-
tomers were encouraged to "get the answers and advice for
your toughest medical ethics dilemmas." By subscribing to
the newsletter, those with one hundred and forty-eight
dollars to spare could, the brochure promised, avail themselves of expertise that might forestall costly lawsuits, avoid
unnecessary conflicts with patients, and minimize the
amount of time spent with various review committees. This
is perhaps a flagrant example of a growing chorus of claims
to moral expertise, but others, both implicit and explicit, are
easy to find in the burgeoning field of applied ethics.

Claims about moral expertise are not manifested only in consulting roles for those who want to pay self-proclaimed moral experts for their advice. Philosophers now routinely sit on committees in medical and scientific settings and have full voting authority with respect to the resolution of problems and the determination of policies. They also have been active participants in commissions sponsored by both public and private organizations that have examined an array of moral issues in medicine. One reason that is sometimes given for allocating philosophers a seat at tables where decisions are made is that a philosopher can represent the interests and views of the lay public. I once sat on a human experimentation committee at Columbia University's Medical Center under the description of layman. In one sense this appellation was entirely apt. I had no idea what the researchers who were my fellow committee members were talking about much of the time. If ignorance is a necessary and sufficient condition of lay status, then, at least upon my initial appointment to the committee, I met that qualification.

But, of course, it is silly to think that philosophers are chosen for committees simply to represent lay interests of one sort or another. At least in the case of my own participation on a hospital human experimentation committee, I was in reality invited and treated as an "expert" in ethics. In general, the justification for according philosophers the privilege of participating in actual decision making seems to rest not on their ability to act as a sort of moral everyman or zealous advocate for the interests of patients and subjects, but rather upon the legitimacy of their claims to moral expertise.

It is not unusual for philosophers (and theologians doing work they describe as applied ethics) to be asked to present testimony before legislative bodies. It is reasonable to assume that the use of philosophers in this capacity results from a growing belief on the part of legislators that philosophers, or at least some philosophers, possess expertise about ethical matters. On the other hand, the explanation might be that philosophers, or at least some philosophers, have successfully hoodwinked others into believing that they possess expertise about ethical matters. In addition, a few philosophers, myself included on one occasion, have appeared in courtrooms as "expert" witnesses. Although such occurrences have been rare, it is interesting to note that some state and local courts have been willing to accept testimony from philosophers about various ethical matters on the grounds that they have expertise that might prove helpful in the resolution of legal conflicts.[1]

Although the use of moral experts in judicial, legislative, and practical decision-making settings does not yet rival the frequency with which social scientists, natural scientists, and various other professionals are utilized for their expertise, the activities of these philosophers merit serious discussion and reflection within and beyond the realm of philosophy. In fact, there has been surprisingly little written about the nature of experts or expertise in ethics. This paucity of discussion is difficult to understand given that claims of expertise and a willingness to adopt the role of moral expert appear to challenge some deeply rooted convictions about the undesirability of talk about ethical expertise and ethical experts. A good deal of the scholarship on Socrates sees him as having little positive to say about those who pretend to possess expertise in ethics.[2] And many modern scholars, such as Karl Popper,[3] have greeted Plato's attempt to defend a form of moral expertise and to create a caste of moral experts to run the state with little more than scorn and derision. Twentieth-century moral theorists have not exactly embraced the notion that their intellectual duties include a commitment to participate publicly in the resolution of either the controversial or the mundane problems of everyday moral life.

Despite the relative silence among philosophers and those in cognate fields about what some working in the area of applied ethics now do, or what is being done in their name by publishers, companies, and advertising agencies, the question of whether either moral experts or moral expertise exist is as germane today as it was in the time of Socrates and Plato. Is it appropriate for any philosopher under any circumstances to claim to be a moral expert? Should legislatures, courts, and other institutions of society seek out those with alleged moral expertise, or are such efforts incompatible with personal responsibility for one's actions and behavior as well as any reasonable theory of democracy?

Quiet Grumbling
Rather Than Written Outrage

Statements about moral expertise or declarations of the status of moral expert have been greeted with more or less stony silence from the pens of philosophers. Although one can find an occasional contemptuous dismissal of the notion that applied ethics rather than metaethics is the appropriate subject matter for ethics courses,[4] little written commentary has been offered by those within the field of ethics on the activities of those described either as moral experts or as possessing moral expertise.

On the other hand, the increasing frequency with which those doing applied ethics have called themselves or have been labeled either as experts or as possessing moral expertise has occasioned a great deal of behind-the-scenes sneering, derision, and contempt. Questions such as, "How does it feel to have to lower your standards of argument when you appear on television?"; "Why do you offer your philosophical services to the agents of repression within society?" (the reference here was to someone doing business ethics); and "Do you do any real philosophy besides medical ethics?" all capture the attitude of many philosophers toward claims of moral expertise, an attitude that suggests a somewhat less than ecstatic current of opinion about declarations of moral expertise.

Some philosophers, including former Secretary of Education, William Bennett, see activities in applied ethics as no more than disguised political activism instigated by the upheavals surrounding Vietnam, Watergate, and the civil rights movements for minorities and women.[5] Others view claims of moral expertise as nothing more than a dubious gimmick utilized by those unable to secure teaching positions.[6] Still others explain the growth of applied ethics as the co-opted response of those facing hard financial times in the humanities to the siren song of power and easy money wielded by the forces of power and repression in bourgeois capitalist states.[7]

It is true that some philosophers view applied ethics with enthusiasm, seeing the recent wave of applied work in philosophy as a refreshing break from the endless series of conceptual thought-experiments and artificial dilemmatic constructs that has passed for moral theorizing for a number of years.[8] However, it is doubtful that even this group is sympathetic either to self-proclamations of expertise or to calls for the creation of standards and certification requirements to protect the practices of a new guild of moral experts who aim to provide succor to the morally bereft and befuddled.

Claims of expertise are not unique to applied ethics, though. Those who teach moral philosophy in a university or professional school surely believe that they possess some sort of expertise that makes it appropriate for them, and not someone from a different discipline or profession, to teach such courses. However, the mores of academia do not encourage expressions of expertise as the basis for one's authority to hold forth in a classroom. Moreover, the subculture of the humanities within universities has little patience with, and no time for, those who would try to demean the profession of philosophy by making it a skill in need of a license, as is the case for psychologists, plumbers, and hairstylists. The expanding activities and influence of those doing applied ethics nevertheless demand that serious attention be paid to a subject many moral philosophers would prefer to avoid or ignore—Is there such a thing as expertise in ethics, and if so, are there moral experts?

What Have Philosophers Thought About Moral Expertise and Attempts to Apply This Expertise by Moral Philosophers?

Though twentieth-century philosophers have written little about whether moral experts exist, most of what has appeared in print has been critical of both the possibility of expertise and claims to expert status. A. J. Ayer, for example, summarily rejects the notion of moral expertise in no uncertain terms:

> ...it is silly, as well as presumptuous, for any one type of philosopher to pose as the champion of virtue. And it is also one reason why many people find moral philosophy an unsatisfying subject. For they mistakenly look to the moral philosopher for guidance.[9]

In the foreword to P. H. Nowell-Smith's widely read introductory text on ethics, Ayer is equally dismissive of moral expertise and experts:

> There is a distinction, which is not always sufficiently
> marked, between the activity of a moralist, who sets out to
> elaborate a moral code, or to encourage its observance, and
> that of a moral philosopher, whose concern is not primarily to
> make moral judgments but to analyze their nature.[10]

One might view Ayer's out-of-hand rejection of moral expertise as reflecting his general skepticism about objectivity in
ethics. But other distinguished philosophers who have been
far more sanguine about the prospects for objectivity in
morals have had few kind words for those who would tout
themselves as moral experts or advance claims about the
privileged possession of moral expertise:

> ...it is no part of the professional business of moral philoso-
> phers to tell people what they ought or ought not to do.... Moral
> philosophers, as such, have no special information, not avail-
> able to the general public, about what is right and what is
> wrong; nor have they any call to undertake those hortatory
> functions which are so adequately performed by clergymen,
> politicians, leader-writers [editorialists]....[11]

Philosophers have tended to confine their doubts about the
possibility of moral expertise and moral experts either to
concise dismissals or to private grousings to those with
similar reservations. Those outside philosophy, who do not
face the restrictions imposed by the norms of collegiality and
professional tolerance that obtain within an academic field,
have not felt similarly constrained, however, and have
greeted such claims with scorn and derision.[12]

Politically Motivated Criticism of Moral Experts and Moral Expertise

Claims about moral expertise have been subjected to
rather vitriolic criticisms on political grounds. From the

right, William Bennett, former Secretary of Education and then head of the National Endowment for the Humanities has written, "One of the most serious ethical problems of our time has become the fad of the new ethics and its defenders."[13] From the left, Cheryl Noble has written:

> To be amenable to these techniques [applying the expertise of the moral philosopher] moral problems must be abstracted from their social settings so that they appear purely moral. In fact, were moral questions more richly defined and conceived, the professional inability of philosophy to deal with them would be obvious.[14]

It is interesting that the political critics of moral expertise do not argue against the utilization of ethical theory in practical affairs or day-to-day matters on the grounds that moral expertise does not exist. None of those motivated by political concerns appears to doubt that, at least in principle, expertise in morals might be possible. What critics such as Bennett and Noble appear to believe is that:

(i) Moral expertise might exist, but moral philosophers are no more likely to possess it than anybody else; and

(ii) Whether he or she has moral expertise, it is not the moral philosopher's job to utilize moral expertise to solve other people's moral problems, or to render judgments about moral matters in the professions or concerning practical affairs. Instead, the moral philosopher should be concerned with promulgating correct modes of thinking and reasoning about ethics to those in need of moral instruction.

Why Are The Philosophical
Critics Skeptical?

It is difficult to know precisely what motivates the distaste many philosophers have felt, and many contemporary philosophers continue to feel, about claims of expertise or calls for the legitimation of moral experts. Most philosophers, at least in modern times, have been willing to identify and acknowledge all sorts of individuals as experts and to recognize many different kinds of claims concerning expertise. But there appears to be something unique to ethics that makes talk of moral experts and expertise grate on the ears of many philosophers.

A combination of factors seems to be responsible for the negative, skeptical reactions of academic philosophers to talk about experts and expertise in ethics.

A. Moral expertise seems incompatible with the virtues of modesty and humility. Socrates set the tone for philosophy's response to those who would hold themselves up as experts in ethics—skepticism combined with irony and cynicism.[15] Claims of expertise appear to violate the profession's own norms concerning the appropriateness of claims about knowledge, whether the subject is morality or something else.

B. Moral expertise seems incompatible with democracy. Moral expertise threatens the standard position on autonomy within liberal democratic theory because recognizing moral expertise seems to cast doubt on the ability of each individual to be his or her own best judge of values. Mill's highly influential arguments defending negative liberty against the inclination of persons and states toward crude paternalism, which he grounded in the right of each person, no matter how befuddled, to direct his or her own life, seem, particularly in the moral realm, incompatible with assertions of moral expertise by those who would call themselves or be called moral experts.

It is not accurate to argue, however, that Mill or a liberalism inspired by Mill has no tolerance for moral expertise. Mill was not concerned about the possibility that some individuals might possess more expertise than others in moral matters. Rather, he was disturbed by the prospect that persons, expert or not, might be allowed to force their judgments of right and wrong, good and bad on others.[16]

A more accurate characterization of the worry felt by some philosophers is that if moral expertise exists, its very existence undermines the possibility of democracy. Why should every citizen have a say in political and value issues if some are more expert than others at these matters? If there is moral expertise, then ought not experts guide society (as Plato suggested long ago) rather than the morally illiterate and backward?

C. Moral expertise must be rejected because moral philosophers know they lack a warranted theory of ethics that can ground all moral beliefs and practices. By vehemently denying the existence of moral expertise and contemptuously rejecting those who call themselves ethical experts, the dirty little secret of ethics—that in the absence of theological foundations all theories of morality appear to lack ultimate warrant—can be kept hidden. Ethics can remain part of the curriculum as long as those who teach it disavow any expertise in the subject. In this way no one can discern that the discipline has no epistemic clothes!

Skeptics, some old-fashioned Marxists, emotivists, and those who desperately wish to replace divine authority with some sort of irrefutable warrant for a comprehensive theory of ethics are prone to reject the possibility of either experts or expertise in ethics. Many in these groups doubt that an objective theory of ethics is possible, and as a result they dismiss claims to moral expertise out-of-hand.[17]

Many of those who believe in the possibility of objectivity in ethics, and who believe that a theory that can warrant moral beliefs and guide moral actions may yet be found, are skeptical of any attempt to do applied ethics in the absence of such a theory. Kai Nielsen, for example, complains that, without an epistemic Archimedean point, applied or practical ethics is meaningless, since the expertise required cannot be grounded in any defensible moral theory.[18] Would-be-experts would best be advised to delay their coronations until such time as metaethically minded philosophers produce a viable, comprehensive theory that can ground normative judgements about particular cases and problems.

D. Moral expertise involves nothing more than being a moderately intelligent person who is adept at logical argumentation and who has the time and the inclination to learn about the history of ethics and to study the facts surrounding a particular moral issue. In this view moral expertise might exist, but it is nothing more than a general set of intellectual skills. Anyone wanting public recognition on the basis of being generally intelligent thus seems to be nothing more than an intellectual fop—at best a person who is narrow-minded, elitist, and snobbish.

E. Moral expertise is dangerous. From the point of view of political and ideological commentators on the right, moral expertise reduces to nothing more than a journey through a boundless conceptual swamp of values clarification. "Real" expertise in ethics requires people to stand up for "good" values (usually meaning "American" values) such as patriotism, respect for life, civic virtue, obedience to proper authority, and so on. Philosophically disposed dispensers of moral expertise are seen as going on and on about the importance of clarifying moral arguments and the need to have good reasons for moral decisions, but such conceptual analysis is not the stuff of which "real" morality is made.

From the left, moral expertise is suspect because it appears to be entirely abstracted from social, economic, and class features of society. Moral expertise, at least as reflected in the writings of many who work in applied ethics, appears to isolate moral issues from the social contexts in which they arise in order to make conceptual analysis possible.[19] Moral expertise thus is bankrupt because it becomes nothing more than an apologia for the norms of the dominant class. Moral experts, as Hegel once warned, are doomed to be nothing more than conceptual handmaidens to the powerful and dominant within a society.

F. Moral philosophers should not allow themselves to stoop to the practical level demanded in so many other intellectual arenas inside and outside the university. Despite the fact that philosophers are usually courteous toward preachers and those giving advice in the newspapers on all manner of topics, most philosophers, following in the modern traditions of British and American philosophy, want to distinguish and distance themselves from the mundane labors of exhortation, character improvement, and positive thinking. Philosophy is seen to be the province of an intelligentsia, not a trade to be pursued in the manner of plumbing, automobile repair, or fishing. Moral experts, if such there be, are advised to hang their shingles out in some other discipline. Philosophy, properly understood as an intellectual's pursuit, has no room for tradesmen, craftsmen, or practical laborers.[20]

Moral expertise is something that many philosophers seem to believe they possess, at least in terms of being qualified to teach the subject or edit journals devoted to ethics. But it is a skill that merits guilt, denial, and embarrassment when mentioned in public. According to this view of the role of philosophy, moral expertise certainly ought not to be seen as a license to present oneself as an expert practitioner outside university departments. Those who claim moral expertise may be quietly tolerated by the philosophical community, in the way that community tolerates self-proclaimed skeptics, solipsists, epistemological anarchists, and other philosophical deviants. But in many quarters it is a tolerance based upon a mix of paternalism, pity, and superiority, not a product of respect for or credulousness about moral expertise.

Does Expertise in Ethics Exist?

I believe that expertise does exist with respect to ethics. I think some moral philosophers have it and some do not. Moreover, I think it entirely possible for someone to acquire moral expertise without having any contact with moral philosophers or any training in moral philosophy. Nor is it the case that training in moral philosophy is more likely to confer expertise about moral matters upon those who have had the benefit of such training as opposed to those who have not.

Some of the resistance by philosophers to public proclamations of moral expertise stems, as noted above, from a mix of views about the epistemic warrant, or lack thereof, for ethical theory. Others find expertise incompatible with commitments to autonomy, democracy, or the norms of humility and tolerance required for the practice of academic philosophy. It seems bizarre, however, to try to save moral philosophy from the sins of hubris or the psychic terror of living without a warranted foundational theory by simply pooh-poohing the notion of expertise in morality. If it is true that moral philosophers have no expertise about anything pertaining to ethics, that they know no more than anyone else about normative and prescriptive matters, then, as R. M. Hare has cogently observed, the time to "shut up shop" has long since passed.[21] If moral philosophy cannot in any way contribute to the resolution of normative and prescriptive matters, it is a field more akin to anthropology than anything else—interesting for the curiosities it reveals, but useless in terms of the logical support it can provide for normative practices and beliefs.

Before rushing out to hang "closed" signs over the doors of moral philosophers' offices, however, it might be useful to question: (a) whether the concept of expertise in ethics has been adequately analyzed; (b) whether there is any logical or epistemological connection between expertise and experts; (c) whether expertise is a status restricted to only a few or open to many and if so, by what means; and (d) whether expertise in ethics consists in anything more than the use of a moral theory to solve a moral problem. The answers that arise from considering these issues may not only shed light on some important questions about the relationship between theory and practice in any domain, but also will, or at least should, make academic philosophers feel less queasy about those who tout themselves as having expertise in moral matters.

What Is Expertise in Ethics?

The question of what constitutes expertise in ethics is more complicated than the available analyses and comments on the subject reveal. One common strategy for trying to understand the nature of expertise in ethics is to see what it is that those who call themselves experts in ethics know or do. But, as Socrates understood, this is not necessarily the best starting point for an examination of the concept.

One peculiar feature of moral experts is that anyone can claim to be one. Like psychics, water diviners, psychotherapists, and any number of other experts, moral experts need only the will and a public forum to claim the mantle of expertise. But this liberal admissions policy ought to cast doubt on the wisdom of drawing any logical connection between expertise and experts. The question of what to do about expertise is one that requires social, legal, ethical, and economic assessments that extend beyond the existence of people claiming to be experts.

It is an open question as to what ought to be done to or with moral expertise if some poor soul is believed to possess it vis-à-vis public policy, practical affairs, or politics. Our culture is enamored of and intimidated by experts of all sorts—revering specialization over generalization while at the same time admiring the flexibility of the generalist. But this should not blind us to the fact that, even if moral expertise exists and if moral experts who have expertise can be found, whether authority, autonomy, or privilege should be granted to these experts remains an open question. The late 19th century struggles between scientifically minded physicians and their other-minded rivals shows that the American public has not always known how to respond appropriately to expertise.[22]

As Mill pointed out in his discussion of paternalism, there may be sound reasons not to elevate anyone to the lofty status of moral expert with its concomitant authority to control and rule the lives of others. But this does not mean that expertise in matters moral does not exist, or that wise and prudent persons should not heed the advice of experts on a variety of subjects, including ethics, in deciding how to live or behave. It means only that the threat of abuse or error is so great that the social role assigned to moral experts or those with moral expertise must be confined to the tasks of exhortation and giving advice. Society should not create an elite of moral experts who have the authority to impose their judgments on others.[23]

What Is Expertise Generally and Who Are The Experts?

Instead of analyzing who moral experts are and what it is that moral experts say and do, it is more useful to try to get some conceptual purchase on the concept of expertise itself. Ordinarily, expertise refers to the possession of a body of knowledge or a set of specialized skills. There is expertise concerning food and wines, French history, and mathematics, for example. There is also expertise concerning baking, detecting art forgeries, surgery, pole vaulting, and pottery making. In other words, expertise is possible, to use a famous distinction of Gilbert Ryle, with respect both to knowing how to do something and to knowing that some set of propositions is true or false.

What is striking about expertise is that, for many skills and areas of knowledge, it is difficult to obtain. People spend a lifetime studying signatures or examining watermarks in order to be able to claim expertise in narrow, restricted domains of inquiry. The years required to become a skilled pole vaulter or gymnast rival the degree of training required to become a physician or pharmacist.

Other forms of expertise are not, however, quite so demanding. The concept of expertise is associated in contemporary American culture with an exclusionary view of human knowledge—the only persons possessing expertise are experts. Expertise in our culture, to count as expertise, requires knowledge or skill that only a few persons can and actually do possess. It is often said that no one can be an expert in everything. By the same reasoning, if there are experts, they cannot be common. Pundits are prone to note that as the range of things to know and do expands, more and more experts will emerge.

One reason it seems odd to say that anyone can have expertise in ethics is that most moral philosophers and a great many psychologists[24] want to argue that each person has the capacity—perhaps a conscience or moral sense—to know and to do what is right. Part of the negative attitude toward the possibility of expertise in ethics is not that it is based upon the suspicion that some do not know more than others, but rather that it is a reaction to the exclusivity associated with moral expertise. If moral expertise exists, only a few of us can have it, and that seems strange and disturbing in a society committed to democratic liberalism, according to which everyone is capable of moral reasoning and moral points of view.[25]

But that our paradigmatic examples of expertise rely on specialized knowledge or the ability to perform certain skills at a high level of accomplishment ought not to obscure the possibility that expertise is not necessarily limited to only a few or a minority of especially gifted persons. Some forms of expertise may be more widely distributed among the population than others. Yet this possibility need not make the particular knowledge or skill any the less a matter of expertise. The existence of a body of knowledge about some subject or about how to perform some skill or activity is sufficient to create the potential for expertise. In many areas of life, lots of people have expertise about a variety of subjects and skills—how to use a telephone, drive a car, mail a letter, exhibit good manners, use a post office, play games, and so on. Other people may not possess expertise with respect to these activities, but most could probably acquire it if they were motivated and taught properly.

This last claim illustrates another ideological headache for those who subscribe to some form of liberal democracy. We like to think that all persons are equal, that drastic or important differences in intellectual ability or dexterity do not exist, and that everyone possesses the key attributes that really matter in life—free will, the ability to love, and so on. The best illustration of the importance of beliefs about human equality is the insistence by advocates on behalf of the elderly, women, and even children that these groups do not lack the skills and capacities possessed by those who control and dominate society, namely, white, middle-aged men.[26] Serious talk of expertise appears to threaten our commitments to both equality and equal opportunity, however. Moral expertise does so in a direct and obvious manner. But does the concept of expertise carry with it, innately or analytically, the conceptual accoutrements of either exclusivity or inequity? Or are these worries that we project onto the concept for reasons that have more to do with our culture and ideology?

We normally do not think of ordinary skills such as using a telephone or simple forms of knowledge such as knowing the streets in one's neighborhood as representing expertise, partly because we often equate expertise with knowledge or skills that the possessor is self-consciously aware of possessing. But we ought not to allow familiarity to breed contempt with respect to descriptions of the vast areas of expertise each person possesses. One need only to spend time with children or persons afflicted with mental diseases and disorders to appreciate how much expertise normal adults have. Nor do we have to accept the claim that expertise is by definition exclusionary. If all human beings knew how to cook Peking duck or were connoisseurs of fine wine, expertise in these matters would not disappear. Experts might not be possible in a world where many people knew a lot about a broad range of subjects or could perform a wide variety of skills and tasks, or in a society that demanded generalists rather than specialists for social or economic reasons, but expertise would still be possible.

Expert and expertise are concepts that have come to be used interchangeably in our increasingly complex, technologically oriented, specialized society. But this is not a conceptual necessity. Nor should we allow our ideological admiration of experts to blind us to the fact that universal expertise about lots of subjects and skills does not make that knowledge and those skills any the worse for wear as matters of expertise.

Singer and Wells' Definition of Moral Expertise

Many of those who are embarrassed about proclamations of moral expertise or who are hostile to such claims on theoretical or political grounds have some basis for their unease. Often claims of moral expertise are highly abstract and not a little pretentious.

Among contemporary devotees of moral expertise, Peter Singer and Diane Wells are two of the few who have tried to give a serious analysis of the concept. But their analysis is unlikely to make skeptics, doubters, and critics go out and retain the services of a moral guru to guide their lives. In brief, their view is that moral expertise involves:

(1) the ability to reason well and logically, to avoid errors in one's own arguments, and to detect

fallacies when they occur in the arguments of
others...;

(2) ...an understanding of the nature of ethics and
meanings of moral concepts...A reasonable knowledge of the major ethical theories, such as utilitarianism, theories of justice and of rights, will also be
useful...; and

(3) ...being well informed about the facts of the
matter under discussion....[27]

I think Singer and Wells are right to claim that the concept
of moral expertise makes sense. Moreover, it is surely valid
to note, as they do later in their discussion, that moral expertise is not confined to moral philosophers, but can be acquired readily by any competent adult with a mind to do so.

Singer and Wells' explication of the concept nevertheless seems inadequate. Their first condition for moral expertise would apply to any type of expertise or, indeed, to any
intellectual enterprise. If this were the only intellectual skill
required, then logicians, not moral philosophers, ought to be
encouraged to take up the field of ethics because they are, in
this view of expertise, most suited for the field.

The second and third conditions show why logicians are
not, on the other hand, especially suited to moral inquiry and
the development of moral expertise. Singer and Wells are
correct in noting that those engaged in moral inquiry or even
moral problem solving need to be acquainted with the "facts
of the matter under discussion." But this phrase does not do
justice to the level of acquaintance required for expertise in
morals. Expertise, as noted earlier, can describe both knowledge about a subject and knowing how to perform a skill or
activity well. In neither case would it do to say that expertise
exists when one has simply a nodding acquaintance with the
facts of a particular case or issue.

The Application of Moral Expertise
Requires More Than The Facts

The notion that the use of moral expertise requires knowing the facts of a particular situation is closely related to a model of decision making that sees the application of moral theory as simply a matter of subsumption and deduction. The model that still predominates in many quarters is that moral theories are composed of axioms, laws, midlevel principles, and initial and boundary conditions. The "facts" are simply fed into this axiomatic hierarchy of rules and lemmas, so that a conclusion can be deductively arrived at by subsuming the facts under the relevant principles.[28] I have dubbed this model of how applied ethics proceeds "engineering ethics."[29] It presumes that:

(i)   Problems are presented to, not found by, philosophers.

(ii) Problem solving is the goal of applied ethics.

(iii) Problem solving is achieved by subsuming the "facts" relevant to a particular problem under a moral theory.

This model of what constitutes a theory in ethics seems to have been inherited from early and mid-20th century discussions in the philosophy of science concerning the nature of theory structure and explanation.[30] There are, however, serious problems with this view.

First, the notion of application utilized is closely wedded to a dubious view of theory. The most extensive analyses of the structure and content of theories have been advanced in philosophy of science. Yet it is not clear that anyone currently working in philosophy of science, or in the history or

sociology of science for that matter, believes that theories, to
be theories, must be composed of axioms, laws, and principles arranged in an axiomatizable hierarchy. Powerful
arguments have been given for the claim that such an
account does not accurately describe all, or, in the eyes of
some, even any theories in science. Among the most telling
criticisms of the axiomatic-deductive view of theory structure and composition has been the point, made by Kuhn,
Lakatos, Toulmin, and Laudan among many others, that
theories, defined as axiomatic hierarchies of principles, are
not the core units of inquiry in science. Often it is necessary
to look at entire research programs, themata, or "strategies"
to understand the evolution of scientific beliefs.[31]

Nor is there any reason to assume that theories, understood as deductive, axiomatic hierarchies, are or must be the
basic units for understanding the evolution of ethical beliefs.
Indeed, few, if any, philosophers who do applied ethics
attempt to locate a single theory and apply it in the mode
prescribed by the older, positivistic conception of theory.
Rather, application often consists in drawing upon comple-
mentary moral theories or individual concepts to achieve
clarification of points. On occasion, those professing expertise in ethics will invoke moral traditions such as rule utilitarianism or virtue-based approaches to morality that are
not, at least in any obvious sense, compatible with one
another. The expertise involved in this approach would
seem to be knowing which tradition or research strategy in
ethics is appropriate for a particular moral domain or problem. For example, it is entirely plausible to argue for a consequentialist moral view in thinking about the ethics of
human experimentation. At the same time, one might adopt
a Kantian perspective in response to questions about
whether a market in organs from living donors ought to be

permitted. The inconsistency would prove fatal to the arguments only if there were no significant differences between the two types of cases.

Utilizing a range of moral theories and traditions drives those committed to a univocal underpinning of all ethical knowledge batty. Yet expertise in both science and ethics appears to consist in part of knowing not only what theory or theories are defensible, or which moral traditions or paradigms are most defensible, but in how to pick and choose among theories and traditions to provide appropriate answers to specific problems or problematics.[32] It should go without saying that expertise also consists of knowing when theory and tradition have nothing to offer.

Unfortunately, moral philosophy, in its quest to replace God with a secure epistemic foundation for moral theory, has lost sight of the fact that very little of morality involves foundational questions. Rather, what is needed and what calls for expertise, and perhaps experts, is the knowledge necessary to individuate, identify, and classify moral issues and problems in order to bring existing moral perspectives—consistent or not—to bear.

It is not only the concept of theory that is suspect in the prevailing model of applied ethics, however. Telling criticisms have also been directed against the idea that "facts" can easily be separated from and identified independently of theories. The notion that a nodding acquaintance with the "facts" is sufficient to permit the exercise of expertise simply buys into a notion of moral facts that is highly suspect. Often expertise consists in recognizing that moral facts are problematic or identifying as relevant facts that are not recognized as such by those seeking help or advice.

Recently some philosophers have argued that analogical reasoning plays a crucial role in scientific theorizing,

particularly in the biological and medical arenas. The use of
paradigmatic cases seems to be a prominent feature of many
such biological accounts.[33] The use of exemplary cases and
analogical reasoning from clear-cut cases also seem to be
important features of the logic of moral reasoning. Case examples and case studies play critical roles in many domains
of moral inquiry. The axiomatic-deductive model of theory
structure fails to capture this key attribute of theorizing in
both science and ethics. The logical relationship of facts to
theories, when paradigmatic exemplars are being used
rather than deduction from theory, may require a different
conception of the kinds of familiarity with the facts that is
necessary than is suggested by the axiomatic-deductive
model of theory structure.

More importantly, an analysis of expertise in terms of
subsuming facts under theories to draw logically warranted
conclusions is far too narrow an account of the various tasks
and responsibilities of those who have moral expertise and
of the way in which that expertise can be brought to bear to
analyze moral problems.[34] Singer and Wells' account of
expertise highlights the ability to solve problems as central
to the possession of moral expertise. But problem solving is
only one of many tasks that those with expertise are required to undertake to make a useful contribution to the
solution of moral problems. Often, in the confused world of
practical affairs, it is not clear exactly what the moral problems that require resolution are. Moral expertise is sought,
not for problem resolution, but for problem individuation
and identification. The kind of familiarity with the facts
requisite for this task is much more complicated than that
suggested by Singer and Wells' analysis. Occasionally, those
involved in practical affairs are not even aware that a moral
problem exists. The task of those with moral expertise may
be to create moral perplexity where none existed. Again, the

depth, comprehension, and degree of knowledge required to conduct moral diagnosis in a particular context is far more demanding than that suggested by Singer and Wells' requirement of general familiarity with the facts in a specific situation. The facts may not be as they appear even to those who request moral help. Those who offer their expertise must be sure that they allow themselves freedom not only in providing advice but, just as importantly, in defining the nature of the problem to be addressed.

Objections to Expertise Reconsidered

If the engineering model of applied ethics is allowed to go unchallenged, many of the criticisms that have been offered by prominent philosophers as well as the fears about moral expertise that are quietly espoused, at least orally if not in print, are impossible to defuse. But if one challenges the assumptions of the engineering model by pointing out that the model gives a false picture of what it is that experts in ethics do, or at least ought to do, then it is possible to answer many of the criticisms that have been directed against both experts and expertise in ethics.

It is simply false to suggest that the professional virtues of modesty and humility are jeopardized by the possibility of moral expertise. If anything, it is those who lay claim to the mantle of expertise who best understand how little is known or understood with certainty about morality. The hubris inherent in contemporary efforts to locate a single foundational theory sufficient for warranting all moral beliefs is at least on a par with the hubris necessary to appear in public as an expert on matters moral. Expertise and experts are not, or need not be, co-optable to the forces of the left or right where ethics is concerned. This can happen only when moral experts allow themselves to operate as crude moral engi-

neers, mindlessly funneling moral quandries and dilemmas
into the abstract apparatus of their favorite theories. Nor is
expertise and its application any less worthy of intellectual
respect than the activities of those who pursue metaethical
investigations by means of thought-experiments, hypothetical scenarios, or counterfactual case examples. Expertise in
ethics demands a close, intimate understanding of the
norms and values that prevail in a given institution or profession. It requires that those who wish to be perceived as
experts know not only the theories and traditions of ethics,
but also the nuances and complexities of the moral life as it
is lived in a hospital, corporation, newsroom, court, or even
legislative body.[35]

The engineering model makes it not only unlikely but
impossible for those avowing expertise in ethics to contribute anything of use to those interested in the analysis of
moral theory and metaethical questions. But if the assumptions of the engineering model are debunked, as they certainly ought to be, then a wide range of contributions can be
seen to flow from practical experience and application to
theory. To mention only a few such contributions, experts
quickly learn that metaethics needs to reexamine the prevailing commitment to foundationalism rooted in a single
theory that dominates so much discussion and debate within
ethics. Similarly, the criteria used to identify and individuate moral "facts" and problems cry out for far more critical
reflection than they have received to date. The role of modes
of reasoning not built on deduction needs to be examined as
well.

Most importantly, it must be appreciated that the recognition of expertise does not consign moral philosophers of
either a theoretical or practical bent to the ranks of totali-
tarian elitists. One can admit the possibility of expertise in
ethics without believing either that expertise is open to only

an elite few or that society should create a social role that accords power and authority to moral experts. It is possible to live in a world where expertise abounds but there are no experts. Experts, oddly enough, are possible only when others decide to grant them status and authority.

Expertise, however, is a different matter. Expertise in ethics, or in any other area of knowledge or skill, is not, in itself, something to be feared or dreaded. Admittedly, there is no reason to believe that expertise exists merely because there are those who declare themselves to possess it. But that is hardly the problem in philosophy, where all claims of ethical expertise are regarded as suspect.

Expertise in ethics appears to consist in knowing moral traditions and theories. It also involves knowing how to apply those theories and traditions in ways that fruitfully contribute to the understanding of moral problems. But, most importantly, ethical expertise involves the ability to identify and recognize moral issues and problems, a skill that may be enhanced by training in ethics, but one that is not by any means restricted to those who have this training or that is beyond the intellectual capacities of those who do not have this training. When the concepts of moral expertise and moral expert are understood in this way, we may be said to live in a world where moral expertise ought to be common but moral experts ought not.

References

[1]Peter G. McAllen and Richard Delgado (1984) "Moral Experts in the Courtroom," *Hastings Center Report* **14**, 27–34.

[2]Gregory Vlastos (1971) "Introduction: The Paradox of Socrates," *The Philosophy of Socrates*, G. Vlastos, ed. (Doubleday, Garden City, NY), pp. 13–21.

[3]Karl Popper (1963) *The Open Society and its Enemies*, vol. 1 (Harper & Row, New York, NY).

[4]Gilbert Harman (1977) *The Nature of Morality* (Oxford University Press, New York, NY), pp. vii–ix.

[5]William Bennett (1980) "Getting Ethics," *Commentary* (December), 62–65.

[6]Samuel Gorovitz (1986) "Baiting Bioethics," *Ethics* **96**, 356–374.

[7]Cheryl Noble (1982) "Ethics and Experts," *Hastings Center Report* **12**, 7–9.

[8]Stephen Toulmin (1982) "How Medicine Saved the Life of Ethics," *Perspectives in Biology and Medicine* **25**, 736–750.

[9]A. J. Ayer (1954) *Philosophical Essays* (Macmillan, London), p. 246.

[10]A. J. Ayer (1954) "Editorial Foreword," *Ethics*, P. H. Nowell-Smith, (Penguin, Baltimore, MD), p. iii.

[11]C. D. Broad (1952) *Ethics and the History of Philosophy* (Routledge and Kegan Paul, London), p. 244.

[12]Gorovitz, *op. cit.*

[13]Bennett, *op. cit.*

[14]Noble, *op. cit.*

[15]Vlastos, *op. cit.*

[16]John Stuart Mill (1961) "On Liberty," *The Philosophy of John Stuart Mill*, Marshall Cohen, ed. (Random House, New York, NY), pp. 185–319.

[17]M. Lilla (1981) "Ethos, 'Ethics' and Public Service," *The Public Interest* **6**, 3–17; P. Drucker (1981) "Ethical Chic," *Forbes* (September), 159–163; and J. P. Euben (1981) "Philosophy and the Professions," *Democracy*, 112–127.

[18]Kai Nielsen (1982) "On Needing a Moral Theory," *Metaphilosophy* **13**, 97–116.

[19]R. Fox and J. Swazey (1984) "Medical Morality is Not Bioethics—Medical Ethics in China and the United States," *Perspectives in Biology and Medicine* **27**, 336–360.

[20]Daniel Callahan, Arthur Caplan, and Bruce Jennings, eds. (1985) *Applying the Humanities* (Plenum, New York, NY).

[21]R. M. Hare (1977) "Medical Ethics: Can the Moral Philosopher Help?" *Philosophical Medical Ethics: Its Nature and Significance*, S. F. Spicker and H. T. Englehardt, Jr., eds. (Reidel, Dordrecht), pp. 47–61.

[22]Paul Starr (1982) *The Social Transformation of American Medicine* (Basic, New York, NY).

[23]Mill, *op. cit.*, chapter IV.

[24]*See*, for example, L. Kohlberg (1984) *The Psychology of Moral Development* (Harper & Row, New York, NY) and J. Rest (1979) *Development in Judging Moral Issues* (University of Minnesota Press, Minneapolis, MN).

[25]Bruce Jennings (1986) "Applied Ethics and the Vocation of Social Science," *New Directions In Ethics,* J. P. DeMarco and R. M. Fox, eds. (Routledge and Kegan Paul, New York, NY), pp. 205–217.

[26]*See*, for example, R. N. Butler (1975) *Why Survive?* (Harper & Row, New York, NY).

[27]P. Singer and D. Wells (1984) *The Reproductive Revolution* (Oxford University Press, Oxford), p. 200.

[28]Michael Bayles, "Moral Theory and Application," (1984) *Social Theory and Practice* 10,47–70; R. M. Fox and J. P. DeMarco (1986) "The Challenge of Applied Ethics," *New Directions in Ethics,* J. P. DeMarco and R. M. Fox, eds. (Routledge and Kegan Paul, New York, NY), pp. 1–18.

[29]Arthur Caplan (1980) "Ethical Engineers Need Not Apply," *Science, Technology and Human Values* 6, 24–32; and "Mechanics on Duty" (1983) *Canadian Journal of Philosophy,* 8, 1–18.

[30]F. Suppe, ed. (1977) *The Structure of Scientific Theories,* 2nd ed. (University of Illinois Press, Urbana, IL).

[31]Arthur Caplan (1984) "Sociobiology as a Strategy in Science," *The Monist* 67, 143–160.

[32]Abraham Edel (1986) "Ethical Theory and Moral Practice: On the Terms of Their Relation," *New Directions in Ethics,* J. P. DeMarco and R. M. Fox, eds. (Routledge and Kegan Paul, New York, NY), pp. 317–335.

[33]K. F. Schaffner (1986) "Exemplary Reasoning about Biological Models and Diseases," *Journal of Medicine and Philosophy* 11,63–80.

[34]Tom Beauchamp (1984) "On Eliminating the Distinction Between Applied Ethics and Ethical Theory," *The Monist* 67, 514–531.

[35]Gorovitz, *op. cit.*

Persons with Moral Expertise and Moral Experts

Wherein Lies the Difference?

Françoise Baylis

Moral philosophers who work in applied ethics are often asked to share with the general public their expert moral opinions on contemporary ethical issues. They are also sometimes called upon to give expert moral advice to health care professionals, to ethics committees, to governmental and/or quasi-governmental advisory committees, and to private institutions. Also, on occasion, their expert moral testimony is sought by the courts. The assumption underlying these and similar requests is that philosophers specialized in applied ethics are repositories of moral expertise and are, therefore, moral experts whose counsel should be heeded. The problem with this assumption, however, is that although practical moral philosophers may have moral expertise—by virtue of their background knowledge of classical moral theories, their studies in ethics and value theory, and their analytical skills—they are not *ipso facto* moral experts. As will be shown, there is an important difference between persons with expertise in moral matters and moral experts.

Moral Expertise and Moral Experts
Without Moral Truth

At the outset, it is important to respond to those who argue that it is silly to speak of moral expertise, or, for that matter, of moral experts because there is no objectivity in ethics (no independent moral truths). Essentially this criticism rests upon the following assumptions: (1) that moral experts are persons with moral expertise, and (2) that moral expertise requires access to moral truths. On the basis of these assumptions, it is argued that because there are no moral truths, there can be neither moral expertise nor moral experts.[1] But why should we assume that moral truth is necessary for there to be moral expertise?

Consider, for example, those fields other than ethics in which claims to expertise do not rest upon claims to "objective knowledge," such as architecture, economics, genetic counseling, or uncontested areas of law. For instance, a person trained in architecture is expected, at minimum, to have good draughtmanship skills, an understanding of function and form in design, and knowledge of the relevant engineering principles. Expertise in architecture draws upon these underlying basics, but what is essential in justifying the claim to expertise is a well developed (possibly inherent) esthetic sense which can inform the requisite skills and knowledge. This esthetic sense clearly is not a function of some "truth" of architecture.

As for the science of economics, those working in this area are presumed to have acquired basic knowledge in mathematics, statistics, and data collection. Expertise in economics exploits this basic knowledge, but further requires a good sense of which problems are important, as well as an ability to discern interesting correlations between the data. These abilities are in some measure dependent upon the prerequisite knowledge, but there is here no question of "truth."

As a final illustration of this point, consider the field of genetic counseling, where, as with ethics, those trained in the area are aware that their recommendations could amount to an imposition of their personal opinions or values. Expertise in genetic counseling rests upon an understanding of genetic principles and an ability to process probabilistic information. But what is important is the "ability to adapt to different types of medical and surgical problems [as well as]...the ability to get on with people...to listen as well as explain."[2] That is, for the person with expertise in genetic counseling, the mastery of certain nondirective counseling skills is crucial in order to convey effectively the relevant probabilistic information and to respond to questions about genetic disorders, recurrence rates, and so on. Thus, once again, "truth" is not the issue with respect to expertise.

There are many areas in which we do not expect "truth" from those with expertise; rather, we expect knowledgeable answers along with good reasons for those answers.[3] Why expect more of those working in ethics? In ethics, knowledge of the major moral theories forms a basis for, or rather underlies, expertise in that someone with such knowledge could draw upon it when confronted with an ethical dilemma. When making moral decisions, however, a person with moral expertise relies upon his/her moral judgment, not upon some notion of moral truth.

This said, we may now consider the prior assumption—that moral experts are persons with moral expertise—in focusing briefly upon the current debate about moral expertise and moral experts.

What Is Moral Expertise?

According to some, moral expertise primarily requires the mastery of certain analytical skills and extensive knowledge of moral concepts, principles, and theories. A major proponent of this view is Peter Singer. Early in the debate he argued that expertise in ethics requires a familiarity with moral concepts, an understanding of the logic of moral argumentation, and ample time to gather relevant information and think about it.[4] More recently, in a book coauthored with Wells, a similar but more thorough account of expertise in ethics is provided, which lists as necessary requirements:

> ...the ability to reason well and logically, to avoid errors in one's own arguments, and to detect fallacies when they occur in the arguments of others...an understanding of the nature of ethics and the meaning of moral concepts...[as well as a] reasonable knowledge of the major ethical theories...[and finally, knowledge of] the facts of the matter under discussion.[5]

Many are sympathetic to such accounts of moral expertise; others, however, maintain that important skills relevant to expertise in ethics are ignored because of the focus on logical analysis and moral theorizing. Caplan, for example, argues that what is missing from the narrow conception of moral expertise is moral diagnosis and moral judgment—the former being the ability to classify moral problems and to identify moral issues not previously discerned, and the latter, the ability to use moral knowledge to see moral issues from different perspectives.[6] These two additional components, in particular the first, lead Caplan to a contrasting view of expertise in ethics, the most recent formulation of which is:

> Expertise in ethics appears to consist in knowing moral traditions and theories. It also involves knowing how to apply those theories and traditions in ways that fruitfully contribute to the understanding of moral problems. But, most importantly, ethical expertise involves the ability to identify and recognize moral issues and problems, a skill that may be enhanced by training in ethics, but one that is not by any

means restricted to those who have this training or that is beyond the intellectual capacities of those who do not have this training.[7]

Caplan recognizes that knowledge of moral theories and analytical skills are important, but he maintains that these are not all important because the ethicist's greatest contribution to a particular debate often is the discovery of problems not previously thought to exist.

All told, then, there are mere differences in degree and emphasis between those who focus on the importance of technical and conceptual analytic skills, and those who insist that there are other important abilities to be considered. That is, no real conflict overshadows the current debate on moral expertise. Not so, however, with respect to the debate on moral experts.

Who Is a Moral Expert?

As noted earlier, a moral expert is commonly thought to be someone with moral expertise. Thus, it is not surprising that many of those who define "moral expert" provide only a descriptive account of the skills and knowledge required for moral expertise. The problem with this, however, is that although one rightly presumes that moral experts have expertise in moral matters, it does not follow that all those with such expertise are moral experts.

Caplan is one of the few who is careful to insist upon the difference between moral expertise and moral experts. In his article, "Mechanics on Duty," Caplan argues:

> The fact that a person is in possession of moral expertise, technically defined *sensu* Singer, does not thereby render such a person 'expert' at moral matters. As Gilbert Ryle insisted,

> there is a sharp difference between 'knowing that' and 'know-
> ing how.' A person may possess all sorts of expertise concern-
> ing a wide variety of subjects but such knowledge in itself does
> not guarantee that the person will be skilled in its applica-
> tion.[8]

The natural reading of this passage would have it that persons with expertise in ethics "know that," whereas experts in
ethics "know how." Curiously, however, this is not the
intended interpretation because Caplan maintains that
expertise in ethics requires *both* "knowing that" and "knowing how."[9] What the above passage does indicate, though, is
that for Caplan moral expertise is a necessary, but not a
sufficient condition for warranting the distinction of moral
expert. Also necessary according to Caplan is social recog-
nition/social sanction. He writes, "Experts, oddly enough,
are possible *only* when those others who would be experts
decide to grant them status and authority [over moral
matters]."[10] Thus, on my understanding, Caplan believes
that moral experts are persons with some degree of moral expertise, defined *sensu* Caplan, who are endowed with moral
authority and elevated to the status of "moral mandarins."

The problem with this distinction between moral expertise and moral experts is that "expertise" is used exclusively
as a decriptive term referencing certain attributes and abilities, and "expert" is used solely as a title of distinction. I
want to suggest an alternative understanding of these two
notions which would recognize the title "person with expertise" (a title of less distinction that that of "expert"), and
which would also regard "expert" as a term descriptive of an
individual's character and competence (a character and
competence which, on the whole, is demonstrably different
from a person with expertise).

Persons with Moral Expertise/Moral Experts

One of the important differences between persons with moral expertise and moral experts is that, whereas the former are skilled and knowledgeable, the latter are *particularly* skilled and knowledgeable; so there is an important question of degree with respect to that which they have in common. This difference of degree cannot be measured in absolute terms because there is no clear or definite demarcation line. Nonetheless, the difference is as significant as that between novice and person with expertise, and as such it needs to be acknowledged.

The second part of my claim regarding the difference between persons with moral expertise and moral experts concerns the unique abilities and attributes of the latter. In particular, moral experts have a different type of knowledge than persons with moral expertise. This point can most easily be explained by appealing to Ryle's distinction between "knowing that" and "knowing how."

In distinguishing the concept of "knowing that" from the concept of "knowing how," Ryle points to the difference between the ability to acquire and retain knowledge and the ability to organize and exploit knowledge. Accordingly, a person claiming expertise in ethics is one who claims knowledge of the major ethical principles, concepts, and theories. An expert, on the other hand, is one who has also acquired the knowledge that is necessary for (1) understanding the major ethical principles, concepts, and theories; (2) appreciating which of these are relevant to a particular debate; and (3) knowing how to apply these in useful and novel ways so as to contribute to the resolution of actual ethical problems. On this point, Ryle is most eloquent:

> A soldier does not become a shrewd general merely by en-
> dorsing the strategic principles of Clausewitz; he must also be
> competent to apply them. Knowing how to apply maxims can-
> not be reduced to, or derived from, the acceptance of those or
> any other maxims.[11]

At this point a caution is in order. By arguing that moral experts must "know how," I do not intend to suggest that moral decision making can be reduced to a matter of deductive proof. That is, I am certainly not advocating "engineering solutions to moral problems." Recall the distinction introduced by Aristotle between *techné* (technical knowledge) and *phronésis,* the reasoning process appropriate to *praxis* ("practical" knowledge). As parsed by Bernstein,

> ...*phronésis* is a form of reasoning that is concerned with
> choice and involves deliberation. It deals with that which is
> variable and about which there can be differing opinions
> (*doxai*). It is a type of reasoning in which there is mediation
> between general principles and a concrete particular situa-
> tion that requires choice and decision. In forming such a judg-
> ment there are no determinate technical rules by which a
> particular can simply be subsumed under that which is gen-
> eral or universal.[12]

The kind of "know how" that I am attributing to the moral expert, therefore, is not technical "know how" whereby moral data are subsumed under a particular moral theory and a conclusion is derived therefrom. Rather, I am advocating practical "know how" whereby the focus of one's inquiry into a particular moral issue is on the different values and the ultimate aims of the various parties to the dilemma.

The last point to stress is that *phronésis,* unlike *techné,* demands understanding of and involvement with other human beings (Aristotle makes this point in discussing the variants of *phronésis*). On my account a moral expert must have certain personal attributes (characterological traits)

that would further his or her understanding of, and interaction with, others. Patience, tolerance, and empathy, for
example, would be of particular importance, along with good
interpersonal, communication, and listening skills. Also
important would be a willingness to consider ethical problems from different perspectives, an ability to identify and to
expose one's professional and personal biases (intellectual
honesty), and a readiness to profit from the work of others.[13]

This enumeration of the skills and personal attributes
required of the moral expert no doubt will strike some as
unusual, particularly because moral philosophers are presumed to be *the* moral experts, yet these abilities and characterological traits are certainly not indigenous to philosophy.
But recall that we are not concerned here with philosopher
kings spinning abstract metaethical theories in their dusty,
ivory towers, but rather with persons (perhaps philosophers) doing work in *applied* ethics. Such persons generally
deal directly with real people who have real problems that
demand resolution. Given this, it is foolish to ignore the importance of these types of attributes and abilities. For only
having mastered these will the person working in applied
ethics avoid churning out impersonal, detached moral
appraisals. And as for those who work in applied ethics but
who do not deal directly with persons involved in moral
dilemmas, the characterological traits noted above are still
important, although some to a lesser degree (e.g., the interpersonal, communication, and listening skills).

Philosophers Do Have an Advantage

The last question to consider is whether practical moral
philosophers are the persons most likely to acquire expertise
in ethics and to become *the* moral experts *par excellence*.
First of all, it should be clear that the distinction between

experts and nonexperts, or between expertise and the lack thereof, is not equivalent to the distinction between applied ethicists and the rest of mankind. Membership in a particular intellectual community does not automatically confer upon one the distinction of expert, nor does it necessarily justify a claim to expertise. More appropriately, the distinction, in each instance, should be between those who are skilled and knowledgeable and those who are less so; and this necessarily cuts across professional classes.

Secondly, I think it is important to recognize, as Singer does, that those trained in philosophy have certain advantages over others. The fact is that the likelihood of acquiring expertise in moral matters or becoming a moral expert is undeniably increased by general training in philosophy, given that moral experts require, among other things, the ability to engage in both technical and conceptual analysis. Hence, although others could well attain expertise in ethics, and even merit the distinction of being a moral expert, those trained in philosophy are certainly advantaged. It is worth noting, also, that those trained in ethics represent a self-selected group of people interested in normative issues, and thus it should not be surprising that they, more so than others, would be the persons with moral expertise or the moral experts.

The difference between a person with expertise in applied ethics and an expert applied ethicist should now be clear. The former is someone who has mastered certain analytical skills and who knows the major ethical principles, concepts, and theories. The latter is someone who has developed these same skills to a finer degree—who has a greater command of the requisite knowledge and, furthermore, can use this knowledge to contribute in innovative ways to the resolution of ethical dilemmas, an ability owing in part to his or her personal attributes and to a certain "moral imaginativeness."[14]

Acknowledgments

I am indebted to B. Freedman for his many invaluable comments on the drafts of this paper. I would also like to thank M. Bolton and T. Kyle.

References

[1]A recent statement of this argument can be found in M. Warnock (1985) *A Question of Life: TheWarnock Report on Human Fertilisation and Embryology*, Basil Blackwell, Oxford, p. 96.

[2]B. Fitzsimmons (1985) "Counseling for the Future," *Nursing Times* **81,** 24.

[3]On my account, values constitute good reasons. This claim obviously needs to be defended; however, the required discussion is beyond the scope of this paper.

[4]P. Singer (1972) "Moral Experts," *Analysis* **32,** 116,117.

[5]P. Singer and D. Wells (1984) *The Reproduction Revolution: New Ways of Making Babies*, Oxford University Press, Oxford, p. 200.

[6]A. Caplan (1982) "Mechanics on Duty: The Limitations of a Technical Definition of Moral Expertise for Work in Applied Ethics," *Canadian Journal of Philosophy* **8,** 13–16.

[7]A. Caplan,"Moral Experts and Moral Expertise: Do Either Exist?," *infra* this volume, pp. 57–85.

[8]A. Caplan, "Mechanics on Duty," *supra*, note 6, p. 12.

[9]A. Caplan, "Moral Experts and Moral Expertise: Do Either Exist?" *infra* this volume, pp. 57–85.

[10]*Ibid.* (italics added).

[11]G. Ryle (1949) *The Concept of Mind,* Barnes and Noble, New York, NY, p. 31.

[12]R. J. Bernstein (1985) *Beyond Objectivism and Relativism: Science, Hermeneutics, and Praxis*, University of Pennsylvania Press, Philadelphia, PA, p. 54.

[13]This listing of the personal attributes required of the moral expert is not intended as exhaustive, but rather as indicative.

[14]B. Szabados (1978) "On 'Moral Expertise'", *Canadian Journal of Philosophy* **6,** 121,122.

Ethical Theory
and Applied Ethics
A Reply to the Skeptics

Ruth Macklin

Skeptical Challenges

Skepticism remains in many quarters about the value of bioethics for resolving moral problems in medicine. That skepticism takes a number of forms. One line of attack focuses on the role of ethical theory, pointing out the inadequacy of theories for providing solutions to practical ethical dilemmas. Another criticism focuses on philosophers and their methodology, arguing that to analyze a problem down to the minutest detail is no substitute for giving an answer.[1] Still another skeptical challenge asks why ethics is needed when laws and policies must be in place anyway to govern the behavior of doctors and hospitals, and to protect the rights of patients.

The first two challenges can be cast in the form of skeptical claims about the relationship between ethical theory and applied ethics:

1. Ethical theories are not especially useful in resolving the issues in bioethics; and

2. The philosopher is not especially qualified to deal with issues in applied ethics.

It would be folly to assert that the enterprise of applied ethics is entirely without problems. But these skeptical claims miss the mark, casting the net too widely in their attempt to discredit the roles ethical theory and philosophers can play. The skeptic who doubts the usefulness of ethical theories for resolving practical issues raises a larger challenge to rational inquiry concerning values. If ethical theories are useless, is it not likely that all attempts at rational analysis and systematic resolution of moral problems are doomed?

The core truth in the first skeptical claim is that no one—philosopher, practitioner, or decision maker in any situation—can resolve practical moral dilemmas by a simple process of taking an ethical theory, applying it directly to a case, and coming up with a single right answer. If the first skeptical claim amounts to nothing more than the rather obvious point that ethical theories do not enable philosophers or moral agents to crank out automatic answers to troubling problems, it would hardly deserve the effort required for a careful examination of the relationship between theoretical and applied ethics. As any beginning student knows, philosophy doesn't supply answers to multiple choice questions. It cannot offer a "how-to" guide to ethical quandries. As Mary Midgley notes, "One of the jobs of philosophy is to find and formulate the rules which underlie sense, the inarticulate patterns by which it works, and to try and deal with their conflicts. The art of doing this used to be called wisdom, but we have got too prissy to use such words today."[2]

Fifteen years ago, when I first entered the fledgling field of bioethics, I wondered whether I or my professional discipline could be of any use in helping to resolve practical problems facing doctors. Even today, when I am called for an ethics consultation or confronted with a case presented at a regular conference in the hospital, I have some lingering doubts about my ability to help. Struggling within myself over how to resolve the dilemma of a disabled, alcoholic, homeless woman who lived in a subway station, made a public nuisance of herself, and who resisted hospitalization despite the efforts of psychiatrists to have her committed to a psychiatric institution, I was unable to resolve the conflict between her right to liberty and and her medical and social "best interest." Contemplating the case of a mentally retarded girl from a Caribbean country visiting relatives in New York with her family, I couldn't achieve a definitive balance between the obligation to respect the parents' religious convictions when they refused recommended brain surgery for their daughter and the duty to override their refusal and perform a high-risk operation to try to save the girl's life. Conducting an internal debate about whether the higher obligation lay in preserving doctor–patient confidentiality, or in informing the employer of an alcoholic train switchman about his impairment, I kept searching for an escape through the horns of that dilemma.

Now, however, I understand much better than when I began my work 15 years ago just what it is I am doubting. Rarely does bioethics offer "one right answer" to a moral dilemma. Almost never can a philosopher arrive on the scene and make unequivocal pronouncements about the right thing to do. Yet despite the fact that it has no magic wand, bioethics is still useful and can go a long way toward resolving the issues, once that phrase is properly interpreted.

A thorough reply to the skeptical claim about the value of ethical theory for applied ethics must defend a series of related points: (a) if a theory is no good in practice, it is not good in theory either; (b) if the difficulty of resolving issues in applied ethics stems from ignorance, disagreements, or uncertainty about the facts, blame should not be laid at the door of ethical theory; (c) ethical theories should not be faulted for the fact that there exists fundamental value disagreements among human beings; and (d) if ethical theories are not especially useful in resolving the issues in applied ethics, then neither is any other rational approach.

My experience as a philosopher working in a medical center has shown that many physicians and other health care personnel faced with hard decisions and ethical dilemmas recognize that there is something we philosophers know how to do that they don't know; or at least, that we can systematically and consistently analyze moral problems in a way they accomplish only occasionally and imperfectly. That ability can be traced more directly to philosophers' competence in constructing well-reasoned arguments and justifications than to their command of ethical theories. Yet, a knowledge of theoretical ethics is also crucial to identifying the source of disagreements and resolving conflicts. Thus, the two skeptical claims are not identical. Claim 1, that ethical theories are not especially useful in resolving the issues in applied ethics, requires a separate response from claim 2, that the philosopher is not especially qualified to deal with issues in applied ethics.

Resolving the Issues

It would be surprising for a philosopher to tackle a controversy without first clarifying its underlying question, or at least offering several interpretations of the meaning of the central claim. True to form, then, I'll begin by analyzing the first skeptical claim. The phrase "resolving the issues" is vague. Additional confusion surrounds the notion of an ethical theory.

The phrase "resolving the issues" can be understood in a strong sense that means "solving moral dilemmas" or "providing clear-cut solutions to ethical quandaries." On this strong interpretation, the skeptical claim asserts that the ability to arrive at correct or uncontroversial moral decisions is not aided by the use of ethical theories. If it is true—and I believe it is—that genuine moral dilemmas have no single right answer, then under this interpretation there is some merit to the skeptics' claim. But this interpretation is not only too strong; it is also too narrow, since only a small percentage of the issues in medical practice and health policy have the characteristics of a genuine moral dilemma. I will argue later that it is expecting too much of an ethical theory to settle controversial moral issues in cases where the controversies themselves can be traced to the disputants' commitment to different and often competing ethical principles. For now, let me simply note that this strong interpretation of "resolving the issues" is neither the only nor the most appropriate way of construing the phrase.

An altogether too weak interpretation of "resolving the issues" should also be rejected. That weak sense sees the philosopher's role as clarifying moral problems and structuring the issues. Although these tasks are part of what bioethicists do in the clinical setting and in their scholarly work, more is involved than mere clarification and structuring. Moral philosophy is not a form of "values clarification," getting people to recognize their own values and discover what they really think about perplexing ethical problems.

A third interpretation lies somewhere between the strong meaning—resolving moral dilemmas—and the too-weak notion of "clarifying the issues." What lies between these unacceptable interpretations? It cannot be captured in a neat phrase, but, roughly stated, it is the ability to produce an argument for one position that is stronger than the argument in favor of a competing alternative. Put another way, when no perfect solution can be found, it is still possible to arrive at the least inadequate or implausible answer to the problem. It is obvious, however, that opponents in a moral debate are not likely to admit that their argument is less plausible or less adequate than their adversary's. Here is where ethical theory enters the picture.

A consideration of the role of ethical theories begins with the reminder that they contain more than simply a recitation of one or more ethical principles or moral rules. An important aspect of any ethical theory—at least any philosophically respectable one—is its epistemological component.

Ethical Theories

Every area of human inquiry has roots in epistemology. Whether the subject is science or ethics, technology, law, or art, questions of what truths exist, how we come to know those truths, and how they can be justified are bound to arise.

Ethical theories embody normative principles, principles that set forth criteria for right and wrong, good and bad, just and unjust actions or policies. Normative principles are the key elements of ethical theories in their application to practical situations. But theories, of morality or otherwise, also contain epistemological features important for the justification of the principles themselves. This is one reason why philosophers, given their background in areas of philosophy besides ethics, are better equipped than most to deal with issues in applied ethics. Although the heart of a moral theory is the fundamental moral principle or principles it embodies, the theory itself consists of much more than its normative content. There is at least as much controversy among philosophers over the foundations of ethics, about the theory of knowledge that lies behind an acceptance of one moral position vs another, and about the meanings of basic ethical concepts, as there is over the selection of the principles themselves.

This point is evident from a moment's reflection on the major differences between teleological and deontological theories of ethics, or between naturalistic and nonnaturalistic approaches. The beginning student of philosophy is often struck by the fact that leading theories are in surprising agreement on the content of most substantive individual judgments. The rather small range of cases in which they appear to disagree—for example, those in which an action that maximizes happiness is in violation of a moral rule—are not only few in number; they are often made to seem more significant by the fabrication of philosophers' examples, such as desert-island cases and fat men stuck in the mouth of a cave in which 10 people are trapped, along with some dynamite. The major differences among Western philosophical theories lie largely in their epistemological underpinnings, and in what they take to be morally relevant factors. Though it is true that these latter differences render ethical theories quite distinct from a philosophical perspective, their distinctness is more often a function of these theoretical differences than of the moral judgments yielded by an application of their normative principles.

The point is that "resolving the issues" in applied ethics should be understood in a way that takes into account the role of the epistemological and conceptual features of ethical theories. There would be no need of ethical theories if the normative principles that lie in the heart of such theories were all that were required. Philosophy students would not have to read the entire text of John Stuart Mill's, *Utilitarianism*; teachers could simply point out the paragraph in which

Mill states the "greatest happiness principle." No one would
have to plow through Immanuel Kant's tortuous *Fundamental Principles of the Metaphysics of Morals*; a half-page
would be enough to list the three formulations of the categorical imperative. And why not just recite the two basic principles arrived at near the end of John Rawls', *A Theory of
Justice*, instead of wading through all 587 pages? A central
aspect of "resolving the issues" in bioethics is attending to
the conceptual, epistemological, and even metaphysical
issues that philosophers deal with as their stock in trade.

   In applied contexts, sometimes the issues are resolved
by getting the participants or disputants to agree on the
morally relevant considerations. Sometimes the issues are
resolved by pointing out that the problem lies not in failure
to assent to the same ethical principle, but rather in disagreement over the empirical facts or probable outcomes of
alternative courses of action. It is a mistake to think that
ethical theories are of no utility, since an important aspect
of applying them is attending to their nonnormative features. It is on this point that replies to the first and second
skeptical claims merge into one: Philosophers are especially
qualified to deal with issues in applied ethics because they
are skilled in addressing conceptual and epistemological
matters as well as knowledgeable about ethical theories.

   In addition to full-blown theories such as those of Kant
and Mill, there is another legitimate category relevant to
responding to the skeptics' challenge: smaller-scale theories
designed to cover a more circumscribed range of actions than
the whole of interpersonal conduct or all social institutions.
A concept that has received much attention in all areas of
applied ethics is paternalism. Both in individual trans-
actions, such as the doctor–patient relationship, and in public policy settings governed by statutes or court decisions, a

wide range of public and philosophical opinion exists about
the justifiability of paternalistic interferences with individual liberty.

Popular sentiment as well as philosophical writings
have tended to be rather critical of paternalism in individual, interpersonal relationships, but somewhat mixed in the
sphere of legal or governmental paternalism, such as bans
on saccharin or laetrile and laws requiring the wearing of
motorcycle helmets or seatbelts. Yet if these debates are to
be more than mere expressions of feeling about the permissibility of coercing individuals for their own good, there must
be a theoretical framework for the discussion. A complete
and fully worked out theory of paternalism would need to
include a view about the nature of persons, and about the
ability of people generally to know their own best interests.
It would require an analysis of the meaning of the "best
interests" doctrine, a discussion of the priority problem (individual liberty vs other values), and an account of the
proper role of the state or community in preserving and
furthering the interests of society as a whole.

Most writings on the subject of paternalism have not
been so ambitious as to encompass all these concerns, yet
philosophers such as H. L. A. Hart, Gerald Dworkin, Tom L.
Beauchamp, Bernard Gert, and Allen Buchanan[3] have con-
structed small-scale theories of paternalism, analyzing the
concept and applying it to a range of issues in law and ethics.
A more comprehensive effort is made by Joel Feinberg in his
recent books, which build on his earlier writings and seek to
develop a full theory of paternalism, as well as an account of
harm and offense to others.[4] In the absence of a theoretical
framework, judgments about the desirability of paternalism
in individual cases or in social policies reveal, at best, adherence to received wisdom or to dogmatically held precepts. A

vague, antipaternalistic sentiment is no substitute, in an argument that calls for a justification, for an appeal to underlying moral principles and related presuppositions.

The advantage of having a theory, as philosophers have argued at length, is that it enables particular judgments to be systematic and well-grounded, instead of ad hoc. To the defender of an ad hoc approach or decision making according to whim, the only reply is the same one given to the skeptic who asks, "Why be rational?" Skeptics who question the need for or the value of ethical theories on the grounds that a systematic or rational approach to individual moral problems is unnecessary or undesirable will probably not be convinced by good reasons of any sort. Their skepticism extends beyond questioning the role of theory in doing applied ethics. If we include under the heading of "ethical theory" small-scale theories of paternalism, autonomy,[5] and privacy,[6] among others, it is hard to see how a justification of moral decisions could even be offered, much less adequately defended, in the absence of ethical theory. Theoretical underpinnings are needed to justify steps at any stage in a moral argument. The burden of proof lies with the skeptic to show how moral reasoning or ethical judgments can take place in a systematic or rational way in the absence of theory.

Problems in Applying Theories

Of all the areas of inquiry philosophers have engaged in over the centuries, none has been intended as more "practical" than ethics. Though always putting forward an ideal of some sort, moral philosophers have nonetheless conceived of their efforts as addressed to issues in the real world. If ethical theories are not especially useful in resolving issues in applied ethics, then moral philosophers have been misguided and misleading others for more than 2000 years. What would it mean to judge that a treatise in moral philosophy is good in theory but not good in practice? If the skeptics are right, that sort of judgment makes sense. Yet philosophers who object to the enterprise of applied ethics are not, for the most part, prepared to abandon the wholly respectable activity of constructing and analyzing ethical theories. If utilitarianism or Kant's ethics—to name only two—cannot be applied at all in real-life settings, why should we take them seriously as ethical theories?

It is, therefore, peculiar to maintain that, in principle, ethical theories are impossible to apply. They are, however, very difficult to apply, for a number of different reasons. There is often a lack of sufficient information about relevant facts or future states of affairs. The probabilities of different possible outcomes may be unknown or only vaguely estimated. The problem of how to formulate a maxim governing an action is formidable. There is the difficulty of determining when the happiness produced by an action outweighs the unhappiness, or which duty should take precedence when two or more conflict. Yet these problems, daunting as they are, have not led philosophers to abandon the study of ethical theory. We think sufficiently well of traditional and contemporary ethical theories to teach them to students, write scholarly treatises and publish articles on them in professional journals, and debate their respective merits on points where they conflict. If the skeptics are right, philosophers should cease misleading students and fooling themselves by paying so much attention to theories that, if useless in practice, are no good in theory either.

What motivates skeptical claims of the sort under discussion? Two considerations in applied contexts are likely to give rise to skepticism about the application of ethical theories. The first consideration is evident in situations when there is an ethical problem or dilemma to be resolved and people disagree about what action to perform.

In many (but surely not all) such cases, the dilemma exists precisely because two different ethical principles underlie the competing judgments about what to do. If a dilemma can be traced to a tension between two incompatible theories, or to competing principles central to those theories, then of course the dilemma cannot be resolved by applying an ethical theory. The problem lies not in the difficulty of application but, alas, in the unsettled metaethical problems of competing theories and disagreement over the foundations of ethics.

The second consideration likely to give rise to skepticism about the application of ethical theories is the sheer difficulty of making the application. Even if all parties to a decision agree on which ethical theory is the correct one to apply, the empirical and methodological difficulties encountered in application are still awesome. We need only recall the stock objections to a relatively straightforward theory, such as utilitarianism, to be reminded of these difficulties.

One such difficulty is our uncertain ability to predict outcomes with any reasonable degree of accuracy. There are problems surrounding measurement: whether interpersonal comparisons are necessary, and if so, how to make them. There are epistemological problems of assigning values to possible outcomes and determining whether individual preferences can be inferred simply from people's behavior. Debates persist over whether pleasures and pains are commensurable and whether types of pleasures and pains can or ought to be weighed somehow.

Some of these difficulties are functions of the stage of development of the social sciences. They reflect the fact that the sciences of psychology and sociology—if they are sciences at all—are in their infancy, with regard to both the state of theory and its application in the spheres of prediction and control. If major difficulties in applied ethics can be traced to the state of the art in the social sciences, it is a mistake to focus blame on ethical theories or on the efforts of philosophers. What stands in the way of being able to apply ethical theory in this range of cases are shortcomings in social science theory and practice.

However, some problems of application, such as the conceptual and epistemological ones raised by utilitarianism, do not stem from the primitive state of social science, but from the philosophical theories themselves. For these problems, the shortcoming lies not so much in efforts to apply the theory, but in failure to work out the details of the theory in the first place. A theory that is impossible or exceedingly difficult to apply is flawed in theory as well.

A factor lending credibility to the skeptics' charge is that moral problems in everyday life rarely have the features of paradigms. Philosophers are notoriously clever at cooking up examples to illustrate their points—a cleverness exceeded only by their ability to think up counterexamples to destroy the claims of their opponents. Matters are fuzzy in real-life settings, though, and case examples do not neatly fit the theoretical precepts. The skeptic must recognize that if these features make it difficult or practically impossible to apply ethical theories, they also make it difficult or impossible to arrive at any rational solution to ethical dilemmas. Decision makers ignorant of ethical theories are not, therefore, doing a better job in their ignorance than they would if they struggled to apply a theory. If it is difficult to arrive at a sound resolution of moral problems by using ethical theories, it is at least as difficult to reach ethical decisions in the absence of any theory whatsoever.

Applying Ethical Theory to Biomedicine

An example of the application of ethical theory to practice is the federal regulations in the US that govern biomedical and behavioral research using human subjects. The overall moral standard embodied in current regulations of the Department of Health and Human Services is a blend of utilitarian and Kantian theories. Current regulations require that every institution engaged in federally funded biomedical and behavioral research on human subjects have an Institutional Review Board (IRB), whose task it is to review research protocols. One charge to the committee is to assess the risk–benefit ratio of all research protocols that come before it. All new drugs and devices must be assessed in this manner, as well as experimental treatments for diseases, new surgical techniques, and a wide variety of interventions that do not stand to benefit the research subjects themselves. In carrying out its function of protecting human subjects, the IRB is supposed to determine whether benefits of the research will, in all likelihood, outweigh the risks of harm to the subjects. The regulations governing research require the IRB to determine that "risks to subjects are reasonable in relation to anticipated benefits, if any, to subjects, and the importance of the knowledge that may reasonably be expected to result."[7]

The moral standard embodied in this statement and in the risk–benefit assessment as a whole is clearly utilitarian. Even if there is some chance that research subjects will be harmed in the course of the study, the overall benefits (to themselves or to others) justify the conduct of the research. The difficulties involved in weighing probable risks of harm against likely benefits, as well as the problems of predicting accurately, are considerable. But on the assumption that such calculations are not wholly fictitious, their use in evaluating the ethics of biomedical and behavioral research is a good example of a public policy that adopts the utilitarian perspective.

But the utilitarian standard of morality is not the only one evident in regulations governing biomedical and behavioral research. These same regulations require that in all instances when human subjects are involved as individuals (in contrast to observations of group behavior), their informed consent to research and treatment does not rest solely on the ethical principle that people deserve to be protected from undue harm. A strong Kantian strain underlies the practice of requiring a patient's or subject's informed consent—a strain expressed in the use of phrases such as "human dignity," "the need to respect personal autonomy," and the individual's "right to self-determination" in the medical setting. The items that must be disclosed in the process of obtaining informed consent are aimed at providing sufficient information to enable subjects to arrive at a rational, informed decision about whether to participate. Potential subjects may also be told that they have the freedom to refuse to participate in the research and to withdraw at any time.

Federal regulations governing research on human subjects thus contain a mix of elements from utilitarian and Kantian ethical theory, but no inconsistency need arise from the use of two different theoretical approaches. The utilitarian balancing required for making risk–benefit assessments applies to one aspect of research practice, intended to protect research subjects from harm. The Kantian elements in the informed consent process appear in a different procedure, one designed to protect the rights of research subjects or patients. Aspects of both consequentialist and deontological approaches to ethics are embodied in this public policy.

Yet despite the fact that these two leading ethical theories are incorporated in this general way into federal regulations, a potential for conflict remains. This conflict has its practical expression in an ongoing debate surrounding the conduct of social science research. Part of that debate has focused on the technical question of whether IRB review is necessary for all or some types of social science research. Another aspect of the quarrel is addressed to the more general issue of whether government regulations ought to apply at all to various types of social science research. The portion of the debate relevant to our concerns has to do with the ethical standards that should govern informed consent to research of this type.

The debate begins with the fact that withholding of information from and even outright deception of subjects has been widespread in social science research. Some outspoken social scientists have argued that their research should be exempt from federal regulations, which were designed for biomedical experiments and have only been applied to social science research as an afterthought. Others contend that it is all right in principle for the regulations to apply, but they should be modified to take into account the experimental design of much social and behavioral research, and the requirements for informed consent should be adjusted accordingly. Their argument is that much research could not be done without some measure of deception, or without concealing the purpose of the study or the identity of the researcher.

Opponents in this debate have two different conceptions of what factors should be morally relevant. Those who argue that the usual requirements for gaining informed consent should be modified or abandoned are appealing to utilitarian considerations. They claim that more benefit than harm results from doing the research, a claim that stands in need of empirical backing if it is to be accepted as a premise in the argument. Contributions to knowledge outweigh any likely harm that could come to research subjects as a result of their having been deceived, they say, and the overall loss to humanity would be much greater if entire lines of research had to be abandoned than if less than fully informed consent were obtained. Furthermore, since all subjects are "debriefed" at the end of the experiment, deception occurs for only a short period of time and the subjects end up being informed. The only factors held to be relevant are the benefits and risks of the research, and the only moral principle thought to be relevant is the principle of utility.

Their opponents, those who argue that the same requirements for informed consent should hold for social and behavioral research as for biomedical research, consider the autonomy of research subjects to be morally relevant. This side adheres to the "respect for persons" moral principle, derived from the Kantian precept that people should never be treated merely as a means to the ends of others. This side in the debate is not moved by risk–benefit calculations that purport to show that little harm is likely to befall subjects who are deceived in an experimental setting. Kantians are never moved by an appeal to consequences, unless they are closet utilitarians. For them, neither the loss of significant contributions to knowledge nor the benefits of possible future applications of the research are factors that should influence policy in this area. Even if no measurable physical or psychological harm is likely to befall research subjects, acts of outright deception, withholding information that might affect a subject's decision to participate, and disguised observation by researchers are violations of trust, privacy, or autonomy. The conclusion of the argument is that the proper criterion for moral rightness is respect for persons, not a balance of benefits over harms. Deception should be ruled out in all forms of research on human subjects.

Does this exercise in applying ethical theory resolve the issues? "No," under the strongest interpretation of "resolve the issues," but "probably yes," under the more charitable interpretation. Although it is quite clear to philosophers who study the issue of deception in social science research that the debate comes down to a conflict between two leading ethical principles, this is far from evident to the principals in the debate. When the opponents themselves recognize that each adheres to a different moral principle, they may very well not budge an inch. But a deeper understanding of the nature of the conflict would emerge.

If skeptics who question the value or usefulness of bioethics still maintain that the issues are unresolved, they are demanding too much of the enterprise. When no single, rationally defensible ethical theory exists, it is misguided to expect that an appeal to theory can solve a conflict that stems from the application of incompatible theories.

As long as the debate between Kantians and utilitarians continues, and as long as the Western political and philosophical tradition embraces both the principles of respect for persons and beneficence, there cannot be a resolution of dilemmas traceable to those competing theoretical approaches. But the inability to make a final determination of which theoretical approach is ultimately "right" does not rule out the prospect for making sound moral judgments in practical contexts, based on one or the other theoretical perspective.

The choice between utilitarian ethics and a deontological moral system rooted in rights and duties is not a choice between one moral and one immoral alternative. Rather, it rests on the commitment to one moral viewpoint instead of another, when both are capable of providing good reasons for acting. Both perspectives stand in opposition to egoistic or selfish approaches, or to a philosophy whose precepts are grounded in privileges of power, wealth, or the authority of technical experts.

Philosophical Analysis in Bioethics

The skeptics' first claim remains problematic if we construe the phrase "resolving the issues" to mean "solving ethical dilemmas" or "providing solutions to moral problems." But the skeptics' second claim, that philosophers are not especially qualified to deal with issues in applied ethics, is easier to rebut. This is because "dealing with issues" involves much more than applying ethical theories directly to concrete cases in order to solve moral problems. Few issues are "purely moral." Philosophical analysis of a more general sort is usually called for in applied ethics. This typically includes providing a conceptual analysis, as well as addressing a range of metaethical considerations: illustrating subtle distinctions between facts and values, clearing up ambiguities in key terms, and pointing out flaws in reasoning. Philosophers are likely to be more skilled than others in these pursuits, since if there is any area of expertise that truly belongs to philosophers, it is mastery of techniques of reasoning and analysis.

A good illustration can be found in the cluster of concerns surrounding the beginning and end of human life. Many opponents in the ongoing abortion controversy have maintained that if only we had an adequate account of when human life begins, or when personhood emerges, we could decide, once and for all, about the morality of abortion and set public policy accordingly. This view rests on the belief that there is a strong link between being a person and being the bearer of rights, especially the right to life. The philosopher Michael Tooley makes the equation explicit in his statement: "...in my usage the sentence 'x is a person' will be synonymous with the sentence 'x has a (serious) moral right to life.'"[8] The belief that there is a strong link between personhood and the right to life is further illustrated by a bill proposed in 1981 by antiabortion legislators in the US Congress, entitled "A Bill to Provide That Human Life Shall Be Deemed to Exist from Conception."⁹

However, a close look at the literature in bioethics reveals there is no consensus at all on the definition of "personhood." Writers who hold a feminist bias take the stance that at no stage of development does a fetus meet criteria of personhood. And writers from those religious traditions opposed to abortion offer a standard of personhood that a zygote, a fertilized egg, can meet. My assessment of why it appears impossible for philosophers, theologians, feminists, and legal scholars to agree on criteria for personhood is that the position they already hold about the morality of abortion is imported into the definition of "human life." This is a sophisticated form of begging the question. Attempts to use the concept of a person as a means of resolving the abortion debate are bound to fail if they rely on the logical fallacy of *petitio principii*.

But the situation is even more complicated. Contributors to the literature surrounding the abortion debate cannot even agree on the relevance and importance of defining "personhood." Four distinct views can be discerned.

One view argues that settling the abortion issue once and for all depends crucially on agreement about whether the fetus is a person and, if so, when in its development personhood begins.

A second view maintains that settling the abortion issue has nothing to do with when personhood begins, since abortion may be morally justified even if it is acknowledged that the fetus is a person from the moment of conception.

A third view holds that whether the fetus is a person is irrelevant to whether it should have legal protection. Concerns about the health of the fetus create pressing policy issues regardless of whether the fetus is granted the status of a person.

The fourth position asserts that since it is impossible to agree on criteria for defining personhood, this issue must be seen as entirely irrelevant to arriving at a solution to the abortion controversy.[10]

Defining "personhood" may seem to many to be "merely a matter of semantics." Yet the question of what should count as a person, or what are the proper criteria for personhood, is a profound metaphysical question. Determining the relevance of that question to the abortion controversy is an epistemological matter. When issues of ethics and social policy are locked in perennial controversy, philosophy can illuminate those issues by showing how conceptual confusions and fallacious reasoning contribute to the problem. Although conducting a philosophical examination of these questions and issues will not succeed in resolving the debate over the morality of abortion, it does provide a deeper understanding of why the controversy remains unresolved and may even be irresolvable.

Similar problems arise for issues concerning the termination of life: euthanasia, the definition of death, and the ethics of withholding or withdrawing life supports. The leading court cases, including those that dealt with Karen Ann Quinlan, Brother Fox, Joseph Saikewicz, John Storar, William Bartling, and Claire Conroy, all revolve around the so-called "right to die." Yet, conceptual confusions have added to ethical uncertainty in a number of these cases.

Initially when the Quinlan case attracted public attention in the mid-1970s, there was a common misconception that if only we had a medically adequate and morally acceptable definition of death, then the ethical problem of removing patients from respirators would be resolved. Some believed that the answers lay in adopting the criterion of brain death, cessation of electrical activity in the brain as measured by EEG, an electroencephalograph. According to this belief, if a brain-death criterion replaced the traditional heart–lung definition of death, then moral decisions would be easier. But a look at the medical facts in the Quinlan case and those that followed reveals that Karen Ann Quinlan would not have been considered dead according to a brain-death criterion, although she was irreversibly comatose, with no hope of returning to a cognitive, sapient state. A much more radical departure from the standard heart–lung criterion would be needed before Karen Ann Quinlan and Brother Fox could have been considered dead while in a "persistent vegetative state."

Can bioethics resolve the issues? The answer is "no" if the question is taken to mean, "Can philosophers supply unequivocal, right answers to all moral dilemmas in medicine?" But for other interpretations, the answer is "yes." When there is no single, correct answer to a substantive moral question, bioethics can still make a contribution. Ethicists can analyze actual or proposed policies and laws and address questions at the "metalevel," such as which moral problems should be candidates for laws and regulations, and which are best left to informal resolution by the concerned parties. Another task at the metalevel is providing criteria of adequacy for a policy or a procedure designed to deal with ethical issues in health care.

In applied ethics, sometimes the issues are resolved by getting the disputants to agree on what are the morally relevant considerations. Doctors who come to recognize the right of patients to participate in health care decisions no longer insist that their sole obligation as physicians is to bring about the best medical outcomes for their patients.

Awareness that the public health model governs medical practice, in addition to the traditional clinical model, leads to an acknowledgment that the physician's duty to keep confidentiality is not absolute. Coming to agree on what are the morally relevant considerations may not directly resolve a particular dilemma, but it shows with precision and clarity just what the residual disagreement is about. In this way, the issues are more fully understood, if not completely resolved.

References

[1]This and other criticisms of applied ethics are made by Cheryl N. Noble (1982) "Ethics and Experts," *Hastings Center Report* 12, 7–9. Replies to Noble's article are given in the same issue by Peter Singer, Jerry Avorn, Daniel Wikler, and Tom L. Beauchamp.

[2]Mary Midgley (1985) "Philosophizing Out in the World," *Social Research* 52, 449.

[3]H. L. A. Hart (1963) *Law, Liberty and Morality,* Stanford University Press, Palo Alto, CA; Gerald Dworkin (1972) "Paternalism," *The Monist* 56, 64–84; Tom L. Beauchamp (1977) "Paternalism and Biobehavioral Control," *The Monist* 60, 62–80; Bernard Gert and Charles Culver (1976) "Paternalistic Behavior," *Philosophy and Public Affairs* 6, 45–57 and (1979) "The Justification of Paternalism,"*Ethics* 89, 199–210; and Allen Buchanan (1978) "Medical Paternalism," *Philosophy and Public Affairs* 7, 370–390.

[4]Joel Feinberg (1971) "Legal Paternalism," *Canadian Journal of Philosophy* 1, 105–124 and his four-volume work, *The Moral Limits of the Criminal Law,* Oxford University Press, New York, NY.

[5]Gerald Dworkin (1976) "Autonomy and Behavior Control," *Hastings Center Report* 6, 23–28; and (1978) "Moral Autonomy," *Morals, Science, and Sociality,* H. Tristram Engelhardt, Jr. and Daniel Callahan (eds.), Institute of Society, Ethics and the Life Sciences, Hastings-on-Hudson, NY, pp. 156–171.

[6]Many examples of small-scale theories of privacy exist in the literature in philosophy, political science, and economics. These pertain mostly to issues surrounding private property and what Rawls refers to as the notion of "private society." A different context is that of privacy and confidentiality in biomedical research and therapy, especially in the

field of human reproduction. An example of such a theory addressed to issues in sex research and sex therapy is Richard Wasserstrom's article (1980), "Issues of Privacy and Confidentiality in Sex Therapy and Sex Research," *Ethical Issues in Sex Therapy and Research*, vol. II, W. H. Masters, V. E. Johnson, R. C. Kolodny, and S. M. Weems (eds.), Little, Brown and Co., Boston, MA, pp. 42–60. A general analysis of the distinction between public and private morality is provided by Daniel Callahan's (1981) "Minimalist Ethics," *Hastings Center Report* **11**, 19–25.

[7]45 CFR 46, *Federal Register* (January 16, 1981).

[8]Michael Tooley (1972) "Abortion and Infanticide,"*Philosophy and Public Affairs* **2**, 37–65.

[9]US Congress, Senate (1981) *A Bill to Provide That Human Life Shall Be Deemed to Exist from Conception*, 97th Congress, 1st session, S.158, Washington, DC.

[10]For the sources of these views and a full discussion of the concept of personhood in the bioethics literature, *see* Ruth Macklin (1983) "Personhood in Bioethics Literature," *Milbank Memorial Fund Quarterly: Health and Society* **61**, 35–57.

Bringing Codes to Newcastle

Ethics for Clinical Ethicists

Benjamin Freedman

Bioethics as a Profession

One should admit to some frank embarrassment when writing about the need to codify the ethical elements of work as a clinical ethicist. In its early years, bioethics itself was concerned about providing a critique of the impedimenta of medical ethics: the existing code of ethics, the Hippocratic Oath, and the posited mentality that perceived ethics as a matter of adherence to well-worn homily. Some of this critique was misguided, revolutionary fervor, but much of it was on the mark; and what I shall have to say may be vulnerable to it.

In addition, the enterprise in which I am inviting you to engage is embarrassing because it is clichéd. What could be more banal than asking ethicists to think about what they are doing? The topic seems the height of self-indulgence, "Let's talk about *me!*"

Embarrassment acknowledged, we should move on. For on the other side of banality is a curious reaction that I perceived in myself, and one that you may have shared: What can be more infuriating than to have your own ethics called

into question? And clichés, of course, attain that status by
repeated verification. The relevant cliché we need to think
about at this time concerns the unshod feet of the cobbler's
children.

The sociological literature delineated a number of elements that 'professions' share, distinguishing them from
vulgar occupations or jobs.[1] A profession possesses a body of
theory that affects practice in some significant fashion. This
theory is sufficiently complicated or esoteric to be relatively
inaccessible to those who have not undergone extensive
formal training and/or apprenticeship. Professional work
therefore depends upon the judgment of the professional to
some extent. It cannot be reduced entirely to rote. The practices with which this theory is concerned are valuable to
society if performed well and of serious concern if performed
poorly. The community generally grants some special privileges to members of the occupation (embryonic profession),
but at the same time, the community makes special demands of these professionals. Important among these is the
requirement that the profession be, in part, self-regulating,
and this demand for self-regulation grows in proportion to
the mysteriousness of professional activity. Perhaps because of its absorbing or ego-syntonic nature, the work of
professionals is thought to create a bond between them,
greater than or at any rate broader than that forthcoming
from common pursuits. The result is the formation of a professional culture, which at first infiltrates activity outside of
the nine-to-five routine and finally blurs the distinction between work and other activities altogether. And, of course,
a profession writes a code of ethics.

The above are to be found in most discussions of the sociology of the professions. Much discussion, of course, has
concerned itself with which of these, and in what degree,
may constitute necessary or sufficient conditions for eleva-

tion to professional status. In addition, epiphenomenal elements may be added to the list. A professional is concerned about malpractice insurance (acquisition or rising rate of); a professional appears in distinguished settings, speaking not merely of fact, but of professional opinion. These distinguished settings include courtrooms and *Time* magazine; in these latter, the professional is further distinguished by appearing with an initial capital letter before the person's name (as in Bioethicist Jones), a dignity once reserved for high-ranking politicians. A professional is entitled to a professional's *per diem* rate and so on. All of these perquisites and forms of recognition have their amusing side and were chosen for that purpose. But each has a serious side as well, reflecting some special status accorded to the profession.

The story of bioethics (and in particular clinical ethics) over the last 10 years in North America has been one of the rapid and accelerating development of these stigmata of professions. The dual theory/practice character of professions has, of course, been tautologically apparent from the beginning. The direct employment of theory to practice has been far more surprising, as has recognition from those outside of bioethics that those engaged in it should have their say. (I am thinking, for example, of those clinical ethicists who write within patients' charts, those who are called to testify as expert witnesses, and those who are consulted as a matter of routine before decisions are taken.)

It is particularly important to note the degree to which those characteristics shared by clinical ethics with other professions diverge from the status of the base liberal arts discipline. That bioethics has pretensions to saying something important about practice is not surprising; that others took these pretensions seriously is. It is hazardous to generalize, let alone extrapolate, from one's own knowledge of such a fluid situation. Nonetheless, I would be very surprised if the following departures from academic philosophy (as one instance of a base discipline) are not exacerbated over the coming years: A routinized requirement that clinical ethicists serve some sort of apprenticeship prior to achieving professional status; the establishment of graduate study programs in clinical ethics, increasingly divorced from philosophy departments; an increased expectation on the part of institutions employing clinical ethicists that they be involved in decision making with respect to patient care and institutional policy, and correspondingly a diminished expectation of activity in independent scholarly work. (In all of these cases, I speak of questions of degree alone.) No jurisdiction has yet, to my knowledge, statutory recognition of clinical ethics *qua* profession, for example, in requiring an ethicist to be included in a legally recognized decision-making process, or by way of grant of professional confidentiality. I doubt that I will be able to say that in ten years' time.

These developments are in many ways flattering to bioethicists and have been sought by them. Yet many of us have extremely negative, or at least highly mixed, feelings about them. The point is, however, that this doesn't matter, for the final determinants of a profession are to be found outside of the profession, in society (though society, of course, takes some account of the factors internal to the profession). It is not the case that bioethics has been hijacked, or is in the process of being hijacked, into professional status. (The moans you hear mingle dismay with ecstasy.) But even if it were, there would be little that bioethicists could do about it.

Many professions, at a comparable stage of development, have concluded that they must adopt a code of ethics and that once a group is a profession, it must so act. The quick philosophic response is that there is no necessity here, natural, logical, or even ethical. That other, different groups have connected professionalism to devising a code of ethical conduct grants no reason for clinical ethics to do likewise; the future need not resemble the past; you cannot move from the "is" of professional status to the "ought" of establishing a code. But quick rejection is as wrong as unthinking acquiescence. It is no coincidence that occupations that have achieved professional status have thereupon devised codes of ethics. The connection is not one of necessity, but neither is it coincidence. The conglomeration of elements that go into professional status independently serve as reasons for developing a code, as we well know, in every case but our own.

Why Should Bioethics Call a Code?

What reasonable purpose can be served by devising and adopting a code of ethics for ethicists? Although adjustments will be necessary because of the special character of the profession, the way to start answering this question is by examining the codes of ethics of others. The Canadian Nurses Association, in developing its code of ethics, agreed with my suggestion of five reasonable goals[2]; they are broadly representative of the reflective literature of codes:

1. A code can identify minimal standards of ethical conduct, define a floor for behavior. By doing so, a code can serve as a basis for charges of professional misconduct and resulting disciplinary action.

2. A code can provide some guidance for those ethical dilemmas in which two opposing courses of conduct have something to be said on their behalf. It may do this by way of identifying relevant values, useful heuristics, and so on, thereby clarifying the choice faced without resolving the issue in the sense of marking off the single right path.

3. A code can serve as a useful tool in professional education, by expressing to prospective professionals the kinds of moral issues faced within practice and the sense of the profession on how these are to be thought about (and, perhaps, resolved).

4. A code can serve to generate needed discussion among professionals concerning the moral dimensions of their work. In this way, a code may feed into as well as reflect the professional culture.

5. The adoption of a code serves to clarify the profession's sense of what may be reasonably expected of a competent and ethical practitioner. If challenged, these expectations may at least serve as the starting point for negotiation. Typically, however, the kind of clarification provided by a code goes unchallenged and helps mold the perceptions and expectations held by other professions (in law, medicine, and so on) of the professional.

The code of ethics of the Canadian Nurses Association was designed to meet these five goals and was justified on their basis. Just to cover all logical bases, I will admit that even granting these points, and admitting that a code would address them, the code is still not ethically justified if an alternative approach *that (inter alia) requires the nonestablishment of a code* would serve better. The burden of proof, however, clearly lies with one who wishes to advance this novel approach.

Can relevantly similar points be made about clinical ethics? I believe that they can, in every instance.

1. There is a floor of ethical conduct for an ethicist as for anyone else. It is, for example, wrong to tell falsehoods; if one does so when acting in a professional capac-

ity, one has violated professional ethics as well. The opportunity and motivation to mislead arise often for the clinical ethicist and are all the more difficult to resist because of the range of rationalizations available to the professional casuist.

2. The clinical ethicist, it may be thought, is equipped by training, practice, and inclination to recognize and think through moral dilemmas. The limited and superficial guidance that may be provided by a code is at best gilding the lily, at worst misleading. For one who trusts (somewhat more than I) the effectiveness of training of clinical ethicists, this might sound persuasive. It ought not. Ethical behavior implicates affect as well as cognition, and the most astute thinker can find that he or she has (conveniently or otherwise) neglected some principle of judgment with respect to his or her own conduct. One possible example: Ethicists are fond of applying to the conduct of others, in controverted circumstances, the "principle of publicity": crudely put, would you be prepared to have the decision with its accompanying justification discussed on *60 Minutes*? Ethicists need to be reminded, at frequent junctures, that the principle applies to their own activities as well. A recent survey indicates that 94% of clinical ethicists would prefer not to keep written records of consultations "because notes could be read by unauthorized persons or used in court proceedings."[3] I do not mean to claim that the principle of publicity is an absolutely reliable heuristic, nor that these responses are strictly in violation of it. But I would certainly have suspicions along these lines if we were speaking of any other profession—Wouldn't you?

3. To the extent that clinical ethics embroils its practitioners in difficult moral situations, not found in

academic base disciplines, and to the extent that its education and training includes nonacademic elements, it is worth flagging the moral facts of life within the profession on behalf of those interested in pursuing it as a career.

4. There should not be any need for a special mechanism to alert ethicists of the special ethical problems they face and to generate discussion and professional interchange. On the other hand, the journals have not been inundating us with these discussions, and many of those that have appeared have been written by authors outside of bioethics; so one wonders. Common sense suggests there might be such a need; *vide* the comment about the beam in one's eye and the mote in another's.

5. The need to clarify the reasonable expectations made of the profession seems to me scarcely to admit of dispute. Some prospective employers appear to be under the impression that a clinical ethicist can ride in like a white knight, with a kit bag full of ethical resolutions, legal counsel, and medical prescriptions. (And some clinical ethicists seem to think they may accommodate these employers.) A banal point, common to virtually all codes of professional ethics, involves delineation of those services that may be provided and prohibits exaggerated representation of efficacy on the part of the professional. Precisely this kind of restriction is needed for a profession in the state of flux that surrounds clinical ethics—for the protection of employers, who should not hire under false pretenses, and for employees, who should not be held responsible to fulfill the hopes and dreams of others.

A phenomenon that may be used to illustrate several of these aims of a code of ethics is the promotional literature distributed on behalf of *Medical Ethics Advisor*. This American monthly newsletter, available "for only $148" (US) annually, promises to supply its subscribers with "the answers and advice" for their "toughest medical ethics questions, including: 'Are you prolonging life or death?' 'Just what is quality of life?' 'Who decides one life is quality and another is not?' " Those subscribing to this publication seal their folly by checking a box that reads, "Yes! I want to ensure my medical ethics decisions are the best ones for my patients, my institution, and myself." For those ill-advised enough to subscribe by March 17, 1986, a special bonus was in store: reprints of some earlier material. The subscriber is assured that "You'll get answers to questions like: 'Do nutrition and hydration constitute medical treatment?' 'What are the economic aspects of continuing care for irreversibly comatose patients?' 'Do living wills do more harm than good?' " (Plus, of course, much, much more.)

Professionals in bioethics reading these publicist's blurbs will see clearly enough that much of the material handled by *Medical Ethics Advisor* involves regulation, law, and economics, rather than ethics itself. But this gazette is intended "for health care professionals and executives," who may not be any more accustomed to distinguishing these than is *Medical Ethics Advisor* and its publicists, themselves. At any event, the overblown claims to provide "answers and expert advice" are distributive over many purely ethical questions included within these blurbs as well.

Imagine that a code of ethics for ethicists had been adopted. A clause prohibiting deceptive advertising would surely be included, as has been routinely the case for other professions. Would this promotional literature, and those associated with it, be in violation of such a code? (Several senior figures in bioethics serve on *MEA*'s editorial advisory board. I have no idea whether they have been involved in the preparation of this promotional literature, nor whether they are even aware of its existence.)

However worthy the periodical itself may be, it appears to me that this form of advertisement falls below a tolerable minimum of ethical behavior. A code containing a clause prohibiting exaggerated representations could serve to focus this debate. Should it be determined that this is deceptive advertising, responsible ethicists would presumably withdraw from this publication unless it agreed to rectify its practices. Should it be felt at the close of discussion that, on the contrary, this advertising was appropriate, the code would have nonetheless been useful in generating needed discussion within the profession of bioethics. Perhaps I am wrong; perhaps ethicists *do* have the "answers and advice" for the "toughest medical ethics questions." In that case, the code would have occasioned my own education regarding bioethics, which is, of course, useful as well.

Each of the goals that the Canadian Nurses Association sought to fulfill is therefore appropriate to clinical ethics as well; and many more examples could be given under each of the categories noted. But the final point to note is that even if one or several of these goals were inapposite, the remaining relevant goals might yield sufficient motivation for establishing a code. Take the last goal of a code—that it serves to establish for those outside of the profession reasonable expectations of conduct—and bear with me as I tell you a story. I believe that it is true, although as you will see I cannot confirm all of its details of my own knowledge. Whether true or not, however, the story may be convincing provided only that it seems plausible.

Code Comfort

One of the earliest experiments in clinical ethics was established in 1976, as the result of a successful grant application by a group of philosophers. A group interview was held as the last step in selecting those philosophers who would be put into the participating institutions as clinical ethicists; as I recall, I was one of about 12 persons interviewed for the four positions.

In the course of the interview, a young pediatrician directed a series of questions to me. What would happen, she wanted to know, if, in the course of my work, I were to discover the existence of some seriously unethical practice? What would I do? I responded, as I recall, by temporizing in all of the usual ways. I would try to learn more of the situation; very likely, it is my beliefs or assumptions of the situation that are mistaken, and I need to get the full story. She rejected this response and persisted: You find that things are just as serious as you had imagined them. What would you do? I said some further, usual things; I would speak to the persons involved to get their perspective and to express my beliefs; I would speak to their supervisor. She wanted to know why I refused to answer her question. You have done all these things, she said, and the situation has not been resolved. Some seriously unethical practice, you believe, is being performed at the institution, and you have followed all institutional channels without getting it changed—what would you do?

I was, naturally, more than a little flustered; she was pressing me quite hard. Faced with the question as posed— and saying again that I could not imagine such a hypothetical being the case at a major urban teaching center—I said that I would then go to the newspapers and generally make as much noise about this as I could in an effort to get external pressure to change what internal pressure hadn't budged. (This, at any rate, is the gist of it—I doubt that I was wholly grammatical by that time.)

That part of the interview I remember vividly. I don't recall whether anyone else was asked the same question, and I certainly think I would remember if anyone had been pressed on it as I was. I doubt anyone else was asked this question, in this way, because I think the answer I gave in the end is the only honorable one that may be given, and I believe that all of the people who were appointed were honorable. I wasn't one of them.

There were a number of people on the selection committee whom I knew. One of them told me later that I was a serious candidate, favored by some, but that I was finally rejected because of the unaltering opposition of that pediatrician, on the basis of the above exchange. This all troubled my informant a lot, enough to tell me about it; but he or she was not going to make an issue about the principle involved. The project was too delicate and important. It already faced too many barriers within the institutions that would be hosting these clinical ethicists, and all had agreed beforehand that the philosophers and the clinicians on the selection committee would have veto power over applications.

What relevance does this episode bear today? The fact is, I believe, that many hospitals indulge, if not sanction, practices that will not stand the ethical light of day—medical students learn proctological procedures by practicing upon institutionalized persons with profound developmental handicaps; residents practice intubation upon patients who have just died, prior to declaring them dead; protocols are approved by institutional review boards and are then violated; and other protocols are prepared by department chairmen and receive ethical approval by a congery of associates. The examples can be distressingly multiplied. The reality of these cases rarely calls for delicate reasoning, fine points about metaethical topology, arguable interpretations of Kant, Bentham, or Paul Ramsey, but just common human decency and a willingness to admit that a point that is in the abstract obvious remains obvious when it applies within your own institution.

Since the time of that interview, many things have changed, all of which serve to sharpen the point. I trust I am not alone in observing that included within the mélange of motivations present in those employing clinical ethicists is, upon occasion, the desire to have a bioethical fig leaf. The idea—the totally erroneous idea—that if some practice has a bioethicist's imprimatur it is all right to proceed is far from common, but it seems to me to have spread with distressing rapidity. The idea itself needs to be resisted, by the clinical ethicists above all. But at the same time, we need to reckon with the idea as well.

A code of ethics would be slender help to the clinical ethicist who wishes to avoid being compromised by his institution. (I presume that this is a subset of the total class of clinical ethicists.) But one professionally engaged in ethics may not need great encouragement and powerful ammunition on behalf of preserving his integrity. In any event, it would be something—and as such, a great deal more than exists currently on behalf of a class of persons who are extremely vulnerable, professionally and economically.

A Code Front Moving in From Canada

I can think of a large number of objections to attempting to devise a code of ethics for this particular institution—and I am sure that there are an equal or greater number of objections that would never occur to me. Some of these objections will be valid; perhaps one or more will present insuperable problems; perhaps one of these will, in addition, be so serious as to vitiate the value of the whole enterprise. But I see no *a priori* reason for thinking that clinical ethics is uniquely so complicated, confusing, or delicate an enterprise that it alone among professions should be without a shared and public understanding of the moral dimensions of its practice.

Nor do I think that clinical ethicists are so clear thinking and saintly a group as to be without need of such a code; nor that they are so vicious and/or close minded as to render one otiose. Clinical ethics is neither above nor below the need for a code.

It might be useful to conclude with another listing, prepared on behalf of the Canadian Nurses Association, of those things that a code of ethics could not be expected to do. A code is as likely to fall prey to exaggerated hopes as to unwarranted cynicism. A code of ethics for nursing, as for ethicists, could not do any of the following:

1. It cannot foresee all contingencies or specify what to do under all imaginable circumstances. Hence, it needs to retain flexibility and even ambiguity.

2. It cannot solve all moral dilemmas or eliminate all moral controversy.

3. It cannot please everyone. In the case of nursing, there were those who demanded that the code include an unqualified prohibition upon abortion and would not accept any code that failed to do so. This kind of problem will always arise. A code cannot dogmatically adopt a moral position with which reasonable professionals differ and cannot be held hostage to a minority's moral agenda. It is the nature of a code that it must face up to, and mitigate the effects of, moral relativism, by expressing a consensus view—much more than a majority, much less than unanimity.

4. It cannot eliminate the need for individual judgment as applied in the individual case. The statements in codes need to be interpreted, with good will and common sense.

5. It cannot be a final statement, but must be reexamined, modified, and clarified over time.

These shortcomings having been noted, the nurses felt that an attempt at drafting a code should be made nonetheless. Had they, or other health care professions, decided against adopting a code of ethics, bioethicists would have been among the first to call them to task. This sounds to me like a topic for a paper on universalizability in ethics.

References

[1]I rely here on Ernest Greenwood (1982) "Attributes of a Profession,"*Moral Responsibility and the Professions,* B. Freedman and B. Baumrin (eds.), Haven, New York, NY, pp. 20–33. Greenwood is representing a tradition of sociological literature on the professions rather than introducing a novel understanding.

[2]*See* Preamble, *Code of Ethics for Nursing* (1985), Canadian Nurses Association, Ottawa, pp. 1–3.

[3]The survey, carried out by the US National Institutes of Health, is reported on in Joyce Bermel's (1985) "Ethics Consultants: A Self-Portrait of Decision Makers," *Hastings Center Report* **15, 2.**

Moral Problems, Moral Inquiry, and Consultation in Clinical Ethics

Terrence F. Ackerman

Is there an intellectual method for resolving the moral problems arising in everyday life, and in what does this method consist? Over the last 20 years, moral philosophers have rekindled their discipline's ancient interest in what Dewey called the "problems of men."[1] Yet controversy persists over whether moral philosophers have a useful contribution to make to the resolution of these issues.

In analyzing the problem of methodology, it has become popular to debunk one approach—deductive application of normative ethical theories to case studies or policy issues.[2] Clearly, there are serious problems that its proponents must address regarding its usefulness as a method of moral problem solving. First, there is no normative ethical theory generally accepted as valid. As a result, there is no place from which our deductions might safely proceed. Second, most statements of basic or foundational moral principles are too vague or uninformative to provide clear guidance amid the ambiguities and perplexities of concrete moral problems.

For example, treating patients as ends in themselves admits of numerous interpretations when responding to their refusals of treatment. Third, if the implications of basic moral principles are not clear cut for specific situations, then their "application" really involves development of their content. However, the deductive approach provides no supplemental method for filling out the content of these principles.

Problems with the deductive application of general moral theories provide deadly ammunition for critics who insist that the "moral expertise" of philosophers consists in skills that have no utility in solving moral problems. Among the most prominent skills of the moral philosopher are, (a) knowledge of moral theories and concepts; and (b) the ability to deductively trace out the implications of these theories for human interaction. Since the role of these skills in solving moral problems is obviously wedded to the deductive approach, the shortcomings of this model also undermine the usefulness of the philosopher's variety of "moral expertise."

However, critics have been less forthcoming in their articulation of a suitable alternative methodology. If the deductive model is bankrupt, how shall we avoid reverting to reliance on moral authority or admission of a paralyzing relativism in matters of moral concern? Although Caplan, for example, has provided a sustained critique of the deductive model, he struggles to find an alternative role for persons trained in philosophical ethics. He suggests two ways in which they might contribute to moral problem solving— by applying their skills in discerning and classifying moral issues and by using their knowledge of ethical theories to view moral problems from a variety of different perspectives.[3] Yet these limited suggestions are not placed in the context of a comprehensive view about the process by which we might resolve practical moral problems.

In the subsequent investigation, I formulate the outlines of a view regarding the methodology we should use to
resolve moral problems. The discussion begins with an analysis of the nature of moral problems. This provides a basis
for understanding the goals and methods appropriate to
moral problem solving, the elements of which are examined
in the second section. In the final section, I employ these
results to clarify the role of the ethicist who provides consultation regarding the clinical care of patients.

Moral Problems

The purposes and methods appropriate to the resolution of a problem depend upon the nature of the problem.
Since there are at least three types of problems that are
sometimes characterized as "moral," I begin with an example of the kind upon which this discussion will focus.

> A seventy-year-old woman was referred to a geriatric reha-
> bilitation unit to gain improved facility in walking and self-
> care, following nerve damage to her lower left leg. As a result
> of this injury, her left foot does not lift properly when she
> attempts to walk, and she must use a leg brace and walker to
> get around. She was brought to the hospital after several epi-
> sodes in which she became too weak to return to her bed after
> going to the bathroom. After diagnostic evaluation, it was
> clear that her inability to get around was only a matter of
> decreased strength and balance.
>   Her daughter reported that she has slowly declined since
> coronary bypass surgery 18 months earlier. She now spends
> most of her time in bed, only walking back and forth to the
> bathroom. Her 76-year-old husband bathes her and assists
> her in dressing. In addition, he does all the shopping, house-
> work, and cooking. He says that these chores are very tiring,
> and he wishes his wife could assist in cooking and cleaning.
> The daughter affirmed that her mother now expects her hus-

band to do all the work, and that her father seems all too willing to submit to her mother's persistent demands.

In the unit, evaluation by the occupational therapist revealed
that she is still sufficiently dexterous and ambulatory to perform household tasks. The physical therapist found that her
leg strength has declined through lack of use, but believed
that her ambulation could be significantly improved. Both
therapists noted that she was completely unwilling to take
any initiative in performing exercises. She constantly com-
plained about being too sore and tired to complete her rehabilitation exercises each day, and she would refuse to discuss the
matter after asking to be returned to her room.

The attending physician found no physical basis for her self-
perceived inability to get around. She simply seemed satisfied
to be entirely dependent on her husband in dealing with her
personal needs. However, tensions are rising in the family.
The daughter made tentative arrangements for her mother"s
entry into a nursing home during her last visit from out-of-
town, even though her mother had said on several occasions
that she wants to return home. The husband seems to be only
passively involved in the decision making. The physician
fears that if the patient is placed in a nursing home, she will
quickly deteriorate into further incapacity.

Is it morally permissible to insist that she complete her daily
rehabilitation exercises? Can the staff force her to talk about
the matter when she asks to be left alone? Can they threaten
to discharge her to a nursing home?[4]

The type of moral problem that is the subject matter of
this case has three distinctive features. First, at stake is the
determination of the conditions we will observe in our interactions with one another under specified circumstances.
Second, diverse states of affairs we cherish (for example,
acting upon our own choices and avoidance of physical incapacitation) might be protected or promoted to a greater or
lesser extent by the adoption of different plans of action.[5]
Third, there is an initial lack of social consensus about what

pattern of interaction should be implemented. Moral problems having these characteristics are the subject matter of
practical ethics.[6]

There are other types of moral problems or issues that
do not share these properties. One involves situations in
which the states of affairs to be realized are coherent and
commonly acknowledged, but there is a failure to implement
an appropriate plan of action. This failure may occur for a
couple of reasons. Persons often engage in immoral conduct,
thereby failing to act in accord with appropriate norms of
behavior. The present case would involve this species of
moral problem if the only obstruction to the woman's rehabilitation were a malicious desire of the daughter to shuffle
her off to a nursing home. Methods for dealing with this
species of moral problem involve the various forms of social
control we use to prevent or alter immoral behavior.
Another reason for failure to implement an appropriate plan
of action is the absence of essential physical and social
resources. For example, the patient's need for physical rehabilitation may go unmet because she lacks the insurance and
personal resources necessary to pay for her hospitalization.
Methods for dealing with this type of situation include the
resources of various social welfare agencies.[7]

A third type of moral problem involves situations in
which we are trying to formulate the statement of principles
that define the general right-making or wrong-making
characteristics of social behavior. In this type of situation,
the pattern of interaction appropriate to particular circumstances is not at issue. Moreover, there is a substantial background consensus about the diverse states of affairs we
should cherish and about the plans of action necessary for
their realization in various circumstances. This consensus

is captured in our "considered judgments."[8] The task is to formulate statements of moral principle that capture these considered judgments. This type of moral problem is the subject matter of general normative ethical theory and is the special bailiwick of moral philosophers.

Moral problems of the first type, and the methods appropriate to their resolution, are the focus of our investigation. In my subsequent discussion, use of the phrase "moral problem" is restricted to moral problems of this kind.

Moral Inquiry

Since the goals and methods appropriate to a particular form of problem solving depend upon the nature of its subject matter, our understanding of moral inquiry must draw upon the practical and social features of moral problems.

Moral problems arise when there is a lack of social consensus about conditions to be observed in our interactions with one another in specific circumstances, resulting from divergence in the state of affairs we cherish. The purpose of moral inquiry is to formulate norms of behavior that evoke shared and stable commitments among the members of the moral community. More specifically, it seeks to identify plans of action for dealing with moral problems that evoke social commitments because they will effectively realize states of affairs cherished by members of the moral community after thorough reflection.[9] Thus, resolution of a moral problem involves a broadening of our shared interests or common moral bonds.

In order to achieve shared norms, moral inquiry must construct plans of action with an eye to protecting or promoting the diverse states of affairs cherished by members of the moral community. Of course, realizing the states of affairs initially cherished by different members of the moral community sometimes is not possible, so revising values will be necessary if a solution is to be secured. For example, it may not be possible both to respect all of the wishes of the elderly rehabilitation patient and to promote her well-being optimally. However, too many discussions in medical ethics are permeated by an "either–or" mentality, which requires a simple setting of priorities on one side or the other. In actual circumstances, plans of action can often be formulated that at least partially acknowledge the diverse states of affairs cherished by different members of the moral community. Moreover, the participants in moral inquiry must revise their values to take fuller account of the moral concerns of others. Since we seek plans of action that evoke a shared social commitment, revision of states of affairs initially cherished is an essential feature in the resolution of moral problems.

In order to achieve shared norms that are also *stable* commitments, moral inquiry must involve thorough reflection. In its absence, tentative social agreements are likely to fall apart as new factors bearing upon the resolution of the problem become apparent. The process of constructing a reflective resolution of a moral problem can fail in several ways. (1) There may be inadequate knowledge about empirical factors that must interact with alternative plans in producing various valued or disvalued states of affairs. For example, forcing the patient to complete her daily rehabilitation exercises may ignore the role of psychosocial factors that contributed to her incapacitation and may continue to operate when she leaves the hospital. (2) Reflection may be inadequate because of a failure to identify alternative plans that might better protect or promote cherished states of affairs than existing alternatives. For example, there may be a way of altering the husband's behavior that promotes the patient's investment in rehabilitative goals without actually violating any of her wishes. (3) Reflection can also

fail because of an incomplete assessment of how alternative
plans will obstruct or enhance cherished states of affairs.
For example, failure to respect the patient's wishes may not
substantially violate her autonomy if its exercise is skewed
by certain impairments in her decision-making capacities.
When moral inquiry fails in one or more of these respects,
plans of action initially implemented may be found wanting.
Thus the achievement of stable, shared commitments requires that they be based on thorough reflection.[10]

The methods appropriate to moral inquiry consist in the
analytic procedures useful in achieving stable and shared
commitments. One component is careful identification of
the various cherished states of affairs that members of the
moral community consider relevant to the choice of a plan of
action. For example, in the case presented, there are numerous states of affairs that might be considered: avoiding
further compromise in the patient's physical capacities;
preserving her relationship with her husband and respecting the roles each has assumed within their relationship;
providing assistance in dealing with her affective impairments (such as extreme dependency and pessimism about
her recovery of functional capacities); and respecting the
stated wishes of the patient. Since this set of values or moral
concerns forms the framework from which development of a
solution must proceed, it must be carefully and completely
articulated.

Another component of moral inquiry involves an understanding of relevant data—those states of affairs that will
interact with the plan of action selected in producing various
consequences. In this situation, the relevant facts include
such matters as the nature of the patient's affective impairments, the reasons for her uncooperativeness, the extent of
her rapport with the staff, the level of functional capacity to
which she can be rehabilitated, the willingness of her hus-

band to assist in the process of rehabilitation, the type of
nursing home environment she might enter, and so forth.
Plans of action are not instituted in an empirical vacuum.
Their consequences reflect their interaction with the various
elements of the situation. If a shared social commitment is
to be sustained, it must have consequences members of the
moral community have anticipated in accepting a plan of
action.

A third aspect of the process of moral inquiry is the identification of alternative ways in which the moral problem
might be resolved. In the present case, there are several
possibilities: discharging the patient unless she quickly
changes her attitude about her daily rehabilitation exercises; forcing her to complete her physical therapy each day;
or allowing her to stop rehabilitation sessions when she desires, but insisting that she discuss the matter with the staff
even when she asks to be left alone. Although this phase of
moral inquiry is frequently overlooked by those who take the
"either–or"' approach to moral problems (either respect her
wishes or treat her paternalistically), this may be an extremely creative phase of the process. Since moral inquiry
seeks to identify a plan of action that takes into account the
diverse states of affairs cherished by members of the moral
community, it is essential that creative alternatives providing more effective realization of these values be explored.

A fourth component of the process of moral inquiry
involves the comparison of how alternative plans of action
will achieve or fail to achieve the states of affairs cherished
by members of the moral community. For example, though
merely accepting the woman's wish to be left alone will
respect her as an autonomous decision maker, it may not
help to preserve the current living arrangement with her
husband. Coercing her into daily completion of her rehabilitation exercises may rebuild her physical strength and agil-

ity, but it may not enhance her long-term ability to remain at home if the patterns of behavior leading to the current functional problems are not addressed. Under the exigencies of clinical circumstances, this phase of moral inquiry involves what Dewey once called "imaginative rehearsal."[11] In imaginiative rehearsal, we project how alternative plans will interact with existing conditions to produce various outcomes, based upon our knowledge of the facts of the situation. Carried to its ideal limit (as it might be for policy matters), this aspect of moral reflection involves the experimental test of plans of action to determine whether they will produce socially acceptable outcomes. This is clearly the most crucial aspect of moral inquiry, since it is from this stage that stable and shared commitments to particular plans of action must emerge.

There are several additional observations that should be made regarding this view of moral inquiry. First, it represents moral inquiry as an inherently social process. This does not mean that we cannot deliberate privately since, as the resident on "night call" can attest, sometimes we must. Rather, it means that when engaging in moral problem solving, we commit ourselves to identifying a plan of action that is capable of evoking a shared commitment. Since moral problems involve an initial lack of social consensus, the plan of action chosen must be supportable by others who would engage in the same process of thorough reflection. This feature of moral inquiry has an exact analog in scientific investigation. Claims regarding correlations of events in nature do not merit the status of knowledge unless different investigators following similar procedures achieve the same results.

Second, this view recognizes an irreducible element of choice in moral matters. Moral inquiry involves the identification of patterns of social interaction that will make secure diverse states of affairs cherished by different members of the moral community. Consequently, the content of what is morally obligatory or permissible depends upon what is cherished. Rachels captures this point succinctly when he writes, "Ethics provides answers about what we ought to do, given that we are the kinds of creatures we are, caring about the things we will care about when we are as reasonable as we can be, living in the sort of circumstances in which we live."[12] There are no categorical imperatives.

Third, this view of the purpose and procedures of moral inquiry does not assure that we will achieve closure on all moral issues. However, lack of complete success in resolving moral problems does not in itself undermine the usefulness of a methodology. Rather, the crucial question concerns which conceptualization of the methods of moral inquiry permits the *most effective* resolution of moral problems. Since the deductive model seems unable to secure closure on any moral issue, the present alternative merits consideration. Moreover, as a moral community, we are not confronted by the stark alternatives of either possessing a method that provides for resolution of all moral issues or admitting conceptual and practical chaos in our social relationships. When reasonable people beg to differ, even after thorough moral reflection, political and legal procedures can be created to assure the orderly protection and promotion of the states of affairs cherished by different members of the moral community. Properly formulated, these procedures themselves may possess moral justification.

Fourth, the present analysis of moral inquiry does assign an important function to general moral theory. On the traditional deductive model, moral principles provide the justificatory basis for more specific norms regulating our social interactions. By contrast, the present view makes the moral justification of a plan of action hinge upon its capacity to evoke a shared and stable social commitment among the members of the moral community. Nevertheless, even without a justificatory role, moral principles still play an important regulative function in the process of moral inquiry. Through the comparison of solutions to various kinds of moral problems, recurrent features of justified plans of action get identified. These recurrent features are formulated in moral principles, such as respect for personal autonomy and beneficence. In dealing with new moral problems, general principles provide guidance regarding the features of plans that must be considered in formulating solutions that protect and promote states of affairs cherished by the moral community. Insofar as they summarize the wisdom of past moral experience, moral principles provide critically important methodological tools in constructing solutions to moral problems.

Similar considerations apply to concepts and distinctions, generated in moral theory, that permit identification of the morally relevant features of problematic situations and of alternative plans of action for resolving them. In the case of the paternalism issue, relevant concepts include the notions of autonomy, competence, decision-making impairment, and harm. Important distinctions include various classifications of the forms of paternalism, including weak and strong paternalism, and soft and hard paternalism.[13] Insofar as these concepts and distinctions permit identification of morally important features of situations, they enhance the thoroughness with which the impact of alternative plans of action upon states of affairs cherished by members of the moral community can be assessed and weighed.

Consultation in Clinical Medicine

Proposals regarding the function of the ethics consultant must begin with a description of the circumstances in which the assistance of an ethics consultant is requested. Such a situation occurs when a clinical decision requires the assessment of nontechnical value factors, and the physician is not able to readily employ the conceptual tools (e.g., moral principles and their constituent concepts), factual data (e.g., information regarding regulations on research), or analytic steps of moral inquiry useful in resolving the problem. The physician is faced by a moral problem and needs help in resolving it. Thus, we may propose that the basic function of the ethics consultant is to *facilitate* the process by which reflective resolution of a moral problem may be achieved.

Identification of the specific ways in which the ethics consultant may facilitate the process of moral inquiry must draw upon our analysis of moral problems and inquiry, as well as our understanding of the knowledge and skills that the consultant brings to the clinical setting. Several major points can be made. First, the consulting ethicist can contribute, in the fashion identified by Caplan, to the classification and diagnosis of the moral problem presented by the situation. As noted earlier, the conceptual framework of a moral problem is set *ab initio* by the diverse states of affairs cherished by different members of the moral community. Moral principles, as summaries of past moral experience, categorize the types of cherished outcomes that satisfactory plans of action must protect or promote. Familiarity with the content of key moral principles enables the consultant to identify the important categories of cherished states of affairs at stake in the resolution of a moral problem. For example, our sample case raises the moral issue of paternalism, involving a conflict between the value assigned to the protection of personal autonomy and promotion of personal well-being. Since investigation of a moral problem proceeds from the framework of problematic values, proper classification of the problem is an essential step in the development of a stable and shared resolution.

Second, the consulting ethicist can provide assistance in identifying alternative plans of action. The ability to comprehensively categorize the diverse cherished states of affairs that form the conceptual framework of the moral problem may enable the consultant to suggest creative plans of action that would satisfy diverse moral concerns. Similarly, knowledge of important concepts and distinctions that bear upon the differential impact of alternative plans of action may lead to the recognition of solutions capable of evoking shared and stable social commitments.

Third, the ethics consultant who is familiar with the results of relevant social science research may promote a clearer understanding of the factual components of a morally problematic situation that affect its resolution. For example, a recent study of refusal of treatment examined the types of reasons for which patients refuse treatment and the ways in which medical staff respond.[14] Knowledge of the factors leading to refusals of treatment, particularly factors capable of alteration through appropriate intervention by physicians, may contribute to analysis of the present case and formulation of a satisfactory solution. Since plans of action must interact with actual conditions in producing valued or disvalued outcomes, the consultant's analysis of the factual components may contribute to the resolution of the moral problem.

Fourth, the ethics consultant has an important role to play in the assessment of how alternative plans of action will promote or impede the realization of cherished states of affairs. As noted earlier, this is the crucial stage of moral inquiry, since it is from this assessment that shared and stable social commitments must emerge. The role of the consulting ethicist in this phase of moral inquiry involves the introduction of moral concepts and distinctions that allow fellow investigators to trace the differential impact of alternative plans of action upon the states of affairs they cherish. For example, respect for the patient's wishes in the case presented may have a differential impact upon the outcome of the situation, depending on whether the patient's wishes represent the results of competent or incompetent decision making. The ethics consultant is able to identify the elements of competent decision making, to explain the relevance of the concept to the present case, and to explore the evidence that the patient possesses or lacks minimal decision-making capacity. Similarly, there is a distinction between paternalistic interventions that promote the cherished activities of the patient and those that promote outcomes not valued by the patient, i.e., the distinction between hard and soft paternalism. This distinction may be relevant to the present case, since paternalistic intervention effective in promoting the patient's rehabilitation and functional independence may contribute to the fulfillment of her desire to remain living at home with her husband. Through the introduction of moral concepts and distinctions, and exploration of their relevance to the current situation, the consulting ethicist contributes to the social process by which plans of action can be ranked in terms of their comparative impact upon states of affairs cherished by members of the moral community.

Finally, the ethics consultant may offer recommendations regarding the proper resolution of a clinical moral problem. This role must be carefully circumscribed. It is not the function of the ethics consultant to deliver "right answers" to the moral quandaries of physicians. This role makes sense only if (a) there are moral truths independent of moral choice, and (b) these truths are able to be discovered and applied to morally problematic situations by philosophers using their special investigative skills. By contrast, we have seen that moral inquiry is a reflective process in which we seek to identify plans of action capable of evoking shared and stable social commitments. Appropriate norms of behavior cannot be identified apart from the input of other members of the moral community. Since justified norms of behavior are socially produced outcomes, the ethics consultant cannot claim to know what is morally right independently of participation by others in the investigative process.

Nevertheless, this conclusion does not preclude the permissibility of making recommendations regarding appropriate solutions to moral problems. What is crucial is that the ethics consultant and the physician understand the status of the recommendation. It is not a claim that such-and-such is the morally required plan of action, independently of the reflective assessment of other members of the moral community. Rather, the recommendation is an hypothesis to the effect that the plan of action is capable of evoking a shared and stable social commitment, because it will effectively realize the states of affairs that members of the moral community will cherish after thorough reflection. In making a recommendation, the consulting ethicist must underscore the fact that it does not merit acceptance independently of the confirmatory reflection of the persons consulted.

Moreover, the ethics consultant may be in a relatively privileged position to identify plans of action capable of evoking shared and stable social commitments. The social process of moral inquiry is reflected in a myriad of academic journals, books, newsletters, government publications, and public discussions devoted to moral issues in medicine. In each context, reflective persons articulate proposals regarding the resolution of moral problems in medicine, eliciting thereby the reflective assessment of other members of the moral community. From this reflective social dialog, tentative solutions that evoke stable and shared commitments often emerge (e.g., the norm requiring informed consent by competent adult patients for participation in biomedical research). The reputable ethicist is presumably current in his or her knowledge of these developments. Recommendations that reflect these tentative results of the social process of moral inquiry bear special consideration in the analysis of current moral problems. Nevertheless, these recommendations remain hypotheses whose acceptance depends upon the confirmatory reflection of other members of the moral community.[15] Thus, the consulting ethicist is not able to deliver "right answers" to moral problems, but may provide informed recommendations for consideration by the physician consulted.

Framework for Further Investigation

A guiding assumption of the foregoing analysis has been that clarification of the role of the philosopher in resolving moral problems in clinical medicine must develop within the framework of a systematic view regarding the nature of moral problems and the structure of moral inquiry. Description of the essential features of moral problems provides a basis for formulating the purposes and procedures of the reflective process by which they may be resolved. In turn, characterization of the goals and methods of moral inquiry permits identification of the ways in which the special skills and knowledge of the philosopher may be used to facilitate the social process of moral reflection.

It is possible that divergent analyses of the essential features of moral problems and moral reflection will produce different conclusions regarding the role of the philosopher in resolving moral problems. However much such analyses may differ, two points seem clear. Without systematic analysis of their role in clinical medicine, moral philosophers will be easy prey for critics who hold unrealistic or inappropriate expectations about their contribution to the moral practice

158

of medicine.[16] Even worse, failure to articulate their role systematically may lead philosophers to misjudge the most significant ways in which their specialized knowledge and investigative skills can optimally contribute to the resolution of moral problems in medicine.

References

[1]The phrase is taken from John Dewey's collection of essays (1946) entitled, *Problems of Men,* Philosophical Library, New York, NY.

[2]Among articles recently critical of the deductive approach are the following: Arthur Caplan (1980) "Ethical Engineers Need Not Apply: The State of Applied Ethics Today," *Science, Technology and Human Values* **6,** 24–32; (1982) "Mechanics on Duty: The Limitations of a Technical Definition of Moral Expertise for Work in Applied Ethics," *Canadian Journal of Philosophy* **VIII,** Supplement, 1–18; and (1983) "Can Applied Ethics Be Effective in Health Care and Should It Strive to Be?", *Ethics* **93,** 311–319; Mark Lilla (1981) "Ethos, 'Ethics' and Public Service," *The Public Interest* **63,** 3–17; Kai Nielsen, (1987) "On Being Skeptical About Applied Ethics," *Clinical Medical Ethics: Exploration and Assessment*, T. Ackerman, G. Graber, C. Reynolds, and D. Thomasma (eds.), University Press of America, Washington DC, pp. 95–115; and Cheryl Noble (June, 1982) "Ethics and Experts," *The Hastings Center Report* **12,** 7–9.

[3]Caplan, "Mechanics on Duty," *supra,* note 2, pp. 13–16.

[4]The case is drawn, in summarized form, from Terrence Ackerman and Carson Strong *Medical Ethics: A Clinical Casebook*, Oxford University Press, New York, NY, (forthcoming).

[5]"States of affairs" is a generic term used to cover the various types of things persons may cherish. These include activities, objects, and forms of interpersonal association.

[6]The term "practical ethics" is preferable to "applied ethics" in describing the systematic normative study of concrete moral problems. The difficulty with the term "applied ethics" is its close connection with the deductive model of moral problem solving. Although the alternative view outlined below identifies several ways in which the results of general moral theory may be used as conceptual tools in resolving concrete moral problems, there are obvious differences between views in the role assigned to general moral theory. Therefore, in discussing proper methodology for the normative study of concrete moral problems, it seems less question-begging to describe this area of study as "practical ethics."

[7]Although philosophers do not usually think of these situations as embodying "moral problems," they are categorized as such in common language. My experience as an ethics consultant provides evidence for this claim. Health professionals sometimes claim that they "have a good moral issue" for me. Upon analysis, it turns out to be one of these types of situations.

[8]Of course, the term "considered judgment" is drawn from Rawls' analysis of the method of reflective equilibrium. *See,* John Rawls (1971) *A Theory of Justice,* The Belknap Press, Cambridge, MA, especially pp. 17–22 and 46–53.

[9]Moral problems may be created by particular situations, or by sets of situations involving conflicts among relevantly similar cherished states of affairs. A "course of action" is a plan for dealing with a moral problem created by a specific situation. A "policy" or "policy option" is a plan for dealing with a set of situations raising the same kind of moral issue, e.g., the issue of paternalism. Discussions in the medical ethics literature usually focus upon issues of policy as characterized here. On my view, policies and courses of action are morally evaluated and justified in a similar fashion. Consequently, I use the phrase "plan of action" to refer generally to policies and courses of action.

[10]In this respect, the results of the present analysis closely parallel the requirements for satisfactory moral problem solving outlined by Howard Brody. *See,* Brody, "Applied Ethics: Don't Change the Subject," *infra* this volume, pp. 181–198.

[11]*See,* John Dewey (1957) *Human Nature and Conduct: An Introduction to Social Psychology,* The Modern Library, New York, NY, pp. 178–186.

[12]James Rachels (June, 1980) "Can Ethics Provide Answers?" *The Hastings Center Report* **10,** 32–40.

[13]"Weak paternalism" involves failure to respect a person's choices for that person's own good because he or she suffers from some serious defect in decision-making capacity. "Strong paternalism" protects the individual even though he or she does not suffer from a serious defect in decision-making capacity. "Soft paternalism" involves paternalistic intervention in which the patient's own values are used to assess benefits and harms. By contrast, "hard paternalism" imposes values that are alien to the patient. For a useful classification of types of paternalism, *see,* James Childress (1982)*Who Should Decide? Paternalism in Health Care,* Oxford University Press, New York, NY, pp. 16–21.

[14]*See,* Paul Appelbaum and Loren Roth (1983) "Patients Who Refuse Treatment in Medical Hospitals," *Journal of the American Medical Association* **250,** 1296–1301.

[15]Of course, the "privileged position" of the ethics consultant can be easily transformed into a myopic viewpoint. For example, the moral philosopher steeped in the libertarian literature may seriously misjudge the current status of the reflective consensus related to the issue of medical paternalism.

[16]Inappropriate expectations play a significant role in Lilla's critique of the role of moral philosophy in professional training. *See,* Mark Lilla, "Ethics, 'Ethos', and Public Service," *supra,* note 2.

Ethical Theory and Applied Ethics

Reflections on Connections

K. Danner Clouser

Introduction

This paper concerns the relationship of ethical theory to applied ethics, especially whether and how they help each other. The topic is beginning to be discussed—as well it might be, since we are nearing the third decade of medical ethics in its most recent resurgence. The relationship of applied ethics and ethical theory has weighed on my mind from the very beginning. I was always working on my own reconciliation, but I seldom talked or wrote about it. It was a second-order worry, a metamatter, and in the midst of the flurry over first-order problems, it seemed a bit divisive and perhaps even elitist to indulge in such deliberations. Nevertheless, I suspected all along that it was probably a more appropriate domain for philosophers than resolving medical ethical dilemmas. Although I have for years assumed a position on the relation of theoretical ethics to applied ethics, I have never presented it explicitly. In other contexts I occasionally made reference to my view and even injected a brief discussion, but these scattered bits and pieces have been

misunderstood by some who have written on the subject. It
therefore seems time to address the question explicitly. I
will, however, proceed in a spirit of inquiry. I will present a
few claims and a perspective primarily to provoke consideration of these metaissues. To this end, I aim to stay on the
level of suggestiveness and to avoid being mired in detail.
(Though I realize that one man's suggestive is another man's
mire.)

Of all people, it will not surprise philosophers to learn
that at least two-thirds of this article must be spent on getting clear about what the question is. Philosophers have
generally not been enamored of bottom lines, preferring to
prowl about in preliminaries and prolegomena. And even
this marginal philosopher, who lives amidst bottom lines, is
no exception.

My plan is to pose the question about the theory and
practice of medical ethics in terms of some of my early experiences. Then I will reflect on the general relationship
between ethical theory and applied ethics. Finally, after
dealing with some of its inherent ambiguities, I will address
what is for me the most difficult aspect of the relationship,
namely, the contribution of applied ethics to ethical theory.

Among the matters I want to avoid is the highroad/low-
road perspective on theory vs application. As interested as
I am in an account of the relationship of theoretical ethics
and applied ethics, I do not want that account distorted by
challenges to and defenses of professional prestige. Why
being a theory worker is generally considered a loftier, more
cerebral calling than an applications worker would be an
interesting study, involving, as I am sure it does, a particular
metaphysical view of the world and man. But I raise the
issue only to have done with it. I trust that none of my remarks are motivated by the desire to diminish theory or to

exalt application. Hard, rigorous, involved, conceptual, and clever work goes on in both arenas.

The Problem

I want to digress from good form just enough to make some personal observations about my first experiences in dealing with medical ethics. Seeing a species of the problem before us in its native habitat should be a more congenial and understandable approach. It amounts to a preanalytic view of the puzzle.

I had come out of an analytic philosophy tradition. I had spent considerable time as student and teacher wondering whether good was a natural quality, whether to say something was good was to say "wow" or to say "I like it," and I thought about what emotions, if any, the ideal observer would or should have and whether "good" was a "grading label" or a "commending word." I used to apologize to my college students for philosophy's inability to help them with real moral problems. I explained that philosophy just does not relate to that level of thought and action, although I do seem to remember always thinking that if someone really got all this theory and metaethics straight, surely the right action would become apparent. I could just never find that linkage between metaethics and decisions about what to do.

Now 18 years later, teaching at a medical school, I seemed to experience the obverse. Here the problems were quite practical. They needed resolution, but there seemed no need of theory. Students, left on their own with moral problems to resolve, proceeded with insight and sensitivity. They articulated their reasons, made nice distinctions, and arrived at credible solutions—all without ever appealing to ethical theory.

As a consultant to many programs teaching medical ethics, I observed courses that insisted on beginning with immersion in ethical theory before confronting the standard array of medical moral problems. This approach either paralyzed the students' moral reflections for the rest of the course or was ignored—the students simply began resolving moral problems afresh as though the discussion of theory had never occurred.

My own practice was to deal with ethical theory at the end of the course. At that point, it functioned not as a narrow prescriptive, but as an after-the-fact accounting of the moves the students had made intuitively throughout the course. Looking at the reasons that had been given, at what overrode what, at exceptions that had been argued for, and so on, an underlying system would emerge, exhibiting some logic and consistency. At that point, some of the deadlocks and enigmas we had encountered could now be redescribed in the language of theory. (That is, we might say, "See, that was a conflict of rights against justice," or "That was a conflict of two moral rules, and there was no argument to support the priority of either," or "That was not a moral obligation at all; pursuit of that good was just his own philosophy of life.") And once a semblance of a system was present, it also became possible to ask in a focused way: "What justifies the system?" "What is its foundation?"

The final, lingering remembrance I will mention is the visiting philosopher. In those early days we would often dream of getting *real* philosophers interested in the field. Good, first-rate philosophers. But on those few occasions when we secured a real philosopher to speak or read, it was awful. They seemed irrelevant and inappropriate, off-the-point and off-the-wall. Their universes of discourse did not engage our universe of discourse.

I indulge these impressionistic recollections to show the realistic context from which the relationship I want to examine grows. The most obvious suggestion that emerges from these incidents is that a gigantic chasm exists between ethical theory and the resolution of real moral problems, that theory has no relevance to the moral decisions that are encountered daily by medicine, and that these daily dilemmas have nothing to say to theory.

On the other hand, perhaps ethical theory is implicit, perhaps it already informs the thoughts of all of us so that we go on working with it, using it, even though we don't explicitly appeal to it. In any case, these experiences and speculations lead quite naturally to the next step in my deliberations, namely, the nature of the relationship between ethical theory and applied ethics.

The Relationship

Many analogies of the theory/practice relationship suggest themselves. They are interesting to pursue as analogies, though not here. For example, *Aeronautics / Flying*: Clearly they are related, but being good and knowledgeable at one does not guarantee being good and knowledgeable at the other. *Psychoanalytic Theory / Patient Therapy:* Theory does seem to guide practice, but it is not clear that it matters which theory you hold; your success with patients may well depend on something unrelated to your theory. *Physics / Engineering / Auto Mechanics:* This injects another link in the chain, but I suspect there are such intermediate links for nearly any "theory/practice" pairing we could find. Physics and engineering seem closer to each other than auto mechanics does to either. Auto mechanics seems more rule and recipe oriented. *Esthetics / Painting* (or perhaps *Investing in Art*): The painter or art dealer may know no art theory,

but have an eye for what is good. Perhaps it is a gestalt recognition, but in any case either could function well without
explicit knowledge of any theory. And the good esthetician
may be bad at both painting and investing. There are many
other such pairings: Theology and behavior; mathematical
theory and applied math; pure geometry and civil engineering; economics and banking.

The idea behind most of the pairings is that one can do
quite well without any knowledge of the theory that in some
sense underlies the activity in question. So in what sense
does the theory "underlie" these activities? It is an important question, but I doubt if one answer can be given for all
the pairings. It is pretty clear that in most, the activity arose
prior to the theory that we now say underlies it. It is not as
though we had the theory—nice, clean, and pristine—and
then by a series of deductions we arrived at the practice. We
bored cannon, made pottery, grew peas, and used aspirin before we had a theoretical account of these practices. Indeed
it was phenomena in and around such practices that stimu-
lated theoretical inquiry; theory grew out of and explained
practice. Theory "underlies" the practice in that it gives, as
it were, an after-the-fact accounting in more generalized
terms and categories. However, theory might lead to certain
refinements of the practices as the implications of the theory
become clear.

It might be argued that moral practice is very different
from the activities I have been describing. The simplest way
of putting the difference would be to say that what I have
described deals with manipulating the natural world, but
moral practice deals with human behavior. So whereas in
the former (i.e., the natural world) theory *describes* the
underlying mechanisms of what is taking place, in the case
of human behavior, theory *prescribes*. That makes for a very

different relationship between theory and practice. There is some truth in this claim, but it does not make as big a difference as we are often led to think.

Let's look at the issue more closely. There are four points worth considering.

1. Like the theories just mentioned, ethical theories begin with the practice of morality. The theoretician must begin with what is commonly taken to be morality, or his analysis would end up being about something else. That is, the domain to be investigated and systematized must be marked off by what is commonly understood to be morality, at least as a first approximation. One who simply "thinks up" a morality misses the mark; his theory has no bite; it is not an account of anything, except perhaps his philosophy of life. We do not think of ethics as inventing new moralities, but of giving a more and more adequate account of the "given" in morality. At least it would seem strange for someone to claim a new discovery in morality—that cheating is perfectly OK, or that deceiving and killing are really moral goods! Although it is a highly debated point, there at least *seems* to be a basic morality from which our theorizing begins and to which our theorizing returns for confirmation. It is that which keeps us talking the same language and understanding the same points. It is that which is the initial meaning of morality. From it we abstract, generalize, and reformulate for rigor, clarity, and consistency. But our theoretical product must arise from and be tested against the clear, paradigm cases of basic morality. The helpfulness and prescriptiveness of the theory would then become apparent in deciding the complicated, unclear, marginal cases.

2. Earlier I noted that students could work very well on moral problems without ever calling theory into play. It is time to re-examine that. Just what do students (naive about

ethical theory) appeal to in order to defend points? Well,
they say things like, "But that would be cheating," or "Not to
tell him would be to deceive him, and that's wrong," or
"Refusing to treat him would be killing." Are these appeals
to theory?

They are certainly appeals to *moral rules*. But are moral
rules part of theory? Moral rules are generalizations, and
they do, in some sense, get "applied," or more precisely,
"acted in accord with." Of course, rules must be further
defended, and that defense would surely be part of theory.
When rules conflict, students do sense that some rules may
have priority over others, but all they seem sure of is that a
rule against taking life is more important than, say, a rule
against deceiving. Occasionally they appeal beyond the
conflicting rules to a principle like "least suffering," but they
make all these moves without ever introducing "ethical
theory."

These observations suggest several questions with respect to the theory/practice relationship. The one I will focus
on is: What constitutes ethical theory? There seems to be a
continuum from metaethics through foundations, principles, and rules. Perhaps ethical theory is to practice as form
is to matter. That is, it might be only a question of more or
less theory or practice because one is never completely without the other. Perhaps theory and practice—instead of being
separated by the chasm I initially suggested—are really
closely intertwined. That is, perhaps theory gets called upon
for every practical move in a way that physics does *not* get
called upon for every move of an auto mechanic. The auto
mechanic uses checklists, not theory. It is not *all* of ethical
theory that is called into consideration, but a chunk of that
continuum—mostly moral rules and whatever one appeals
to in order to resolve conflicts among rules and to recognize
exceptions to rules. Using chunks of theory seems ordinary

to us because it is what we've been doing all our lives in moral deliberation. One may never have been taught ethical theory explicitly, but one surely has had one's fill of moral problems about which one has conscientiously deliberated.

The parts of this continuous theory that do *not* get called upon are the foundations and the metaethical issues. One suspects that these parts are generally what are being referred to as "theory" when "theory" is contrasted with "applied ethics." So the now infamous chasm may really exist between the level of actual moral problems and the level of *foundations* of moral theory. That would explain why each of the two worlds seems to function quite well without its practitioners having much knowledge of the other. Thus, in discussions concerning ethical theory and applied ethics, a major problem may have been that the metaethics, the foundations, and the methodologies comprising principles, basic rules, and what we might call "bridging rules" (which are derived from the basic rules in conjunction with the subject matter of a particular domain such as business or medicine) all got lumped together as ethical theory. (Perhaps the best model for this would be law, with its components of foundations, legal theory, legal principles, statutory law, and case law, and the real cases to which this superstructure is believed to be relevant.)

3. In the context of "theory as continuum," we ought to consider the notion of applied ethics being, as it is often put, "simply applying the theory to moral problems." Except for the "simply" I believe this is basically correct—at least if moral rules are part of the theory. In doing medical ethics we are constantly appealing to rules. Don't kill, don't cheat, don't deceive, don't cause pain, don't break promises. Such rules guide us. They tell us where the moral issues are. They show us where to focus. The vast majority of issues are

resolved by appealing to rules. But to say this is misleading because these are not the issues over which we toil. The issues that become problems for us are precisely those for which an appeal to rules is not enough. Often it is not nearly enough.

Rules might conflict with one another, or the situation might be too complicated to see how rules apply, or we might think an exception to a rule is appropriate, or, as is frequently the case, rules might not cut finely enough for the details of the situation. In any case, it then becomes necessary to concentrate efforts "above" or "below" the rules. By "above" I mean to indicate a metalevel concern, for example, we might question the justification of rules or their individuation. We see if anything would justify a priority ranking of the rules. Or, if an exception to a rule seems in order, we try to see what would justify such an exception. We look for criteria for making exceptions, and ultimately, this might lead us to appeal to such fundamental concepts as rationality or impartiality.

We should note that all this would count as "applying" theory—though certainly not as *simply* applying theory"— because it calls on the structures and deliberations of ethical theory that are, in some sense, already in place. Note also that ethical theory is telling us what is morally relevant, not unlike any scientific theory, which generally tells us what facts are relevant. Through the blooming, buzzing confusion of these situations, it is theory that shows us that there is a moral issue at stake, precisely what it is, the appropriate questions to ask, and the morally relevant facts to pursue. In short, ethical theory tells us what is morally relevant.

4. I said that in problem cases an appeal to rules is seldom enough, so we must concentrate "above" or "below" the rules, and all that I have just said about theory is an explication of the "above" branch. Now I want to examine

what it is to focus on issues "below" the moral rule. This is the area wherein a great deal of conceptual work goes on. I think this is the primary focus for those doing applied ethics, particularly for practitioners in a field (e.g., medicine) who are concerned with moral issues. Most articles and discussions concentrate on this conceptual domain.

I have at times described this task—too briefly, to be sure—as "preparing the ground for application of the rules." In the typical problematic moral issue, the trouble lies here, and theory is of little help. The relevant rules may be clear enough and not in conflict; what is not clear is how they apply. For example, we may be sure that deception is immoral, but not sure how much a consent form for treatment must tell in order not to be deceptive. Must it relate every known biochemical process that will be going on in the body? In how fine a detail? Organ level, cellular level, molecular level? Is it deception not to mention alternative treatment possibilities? How many and which? Should faith-healing, chiropractic, herbal, homeopathic, reflexology, and pow-wow be mentioned as alternatives? Is it deception not to mention the surgeon's personal success rate on the consent form? Is it deception not to mention his failures?

Or, to change the rule in question, breaking a promise is certainly wrong, but was there really a promise of confidentiality when the patient consulted the physician? Is letting the secretary type up the notes from that consultation a breach of confidentiality? How about discussing the case with a colleague? Is that a breach of confidentiality and hence the breaking of a promise?

Killing is unquestionably wrong. But is "not saving a life" the same as killing? If it depends on the circumstances, precisely what are the circumstances? Not neglecting one's duty may be a rule handed to us by theory, but what determines one's duty? Has the physician a duty to be available to a patient 24 hours a day, seven days a week? Has the doctor a duty to maintain a patient's life, even if the means are very painful and there's little or no chance of successful outcome? If so, is that duty to the patient or to society?

There are many examples of this line of conceptual exploration. That is why people doing medical ethics can be so busy without ever raising questions about ethical theory. Very many facts must be understood, in considerable subtlety, about practices, understandings, and implicit agreements, as well as diseases and therapies. Additionally, a lot of conceptual analysis is necessary to do the job. Moral matters frequently turn on the understanding of a certain key concept such as "disease," "autonomy," "paternalism," "consent," "informed," "death," "duty," "competence," "manipulation," "causing vs allowing," and so on.

I have been describing a domain of activity that I originally referred to as "below" the moral rule or as "ground preparation." It is central to the field of medical ethics. Inasmuch as it seems ethical theory is not explicitly involved, one may not want to call it "doing ethics." I generally have not done so, for the simple reason that I hated to anger real philosophers. (There's nothing worse than an angry philosopher—especially a real one.) But whatever one calls it, I do think it is a field of endeavor especially amenable to philosophical skills. Moreover, the endeavor is necessary in order to make and justify moral judgments in the practical realm.

The Contributions of Applied Ethics
to Ethical Theory

We have been examining the relationship between ethical theory and applied ethics. Whereas many have assumed that ethical theory does all the work and that applied ethics simply plugs in the values of variables and solves the equation, I have suggested that such a formulation vastly over-simplifies what really goes on. In fact there may be a lot of hard work with which theory does not help at all.

We now turn to the reverse of that issue, namely, whether applied ethics contributes something to ethical theory. But first there are several ambiguities about the issue that must be addressed.

Ambiguities in the Question Itself

1. How would a contribution to ethical theory ever be identified? Would we know one if we saw one? In order to label something a "contribution," would an ethical theorist have to affirm that he or she discovered it while doing applied ethics? Or, perhaps, having had it reported by an applied ethicist, would a theorist have to testify that it led him to change his theory? Fat chance. If that is what constitutes a contribution, we will not have much to talk about.

2. But perhaps that is a bit harsh. Many who do applied ethics are also theoreticians, and I suspect they would say applied ethics has contributed to their understanding and formulation of ethical theory. But then the question becomes: Was applied ethics *essential* to that new understanding and reformulation, or simply the occasion for the new insight? For example, couldn't a theoretician have simply imagined the real-life situation that occasioned the discovery and consequent contribution to theory? Or does applied ethics yield contributions that could have been reaped in no other way?

3. To press the point a bit more, what is the difference between the real moral problems encountered by applied ethics and the hypotheticals posed by ethical theory? Hasn't the theoretician been doing applied ethics all along—at least

hypothetically? The theoretician gives examples, tries out rules and principles on realistic situations, and makes changes or draws conclusions accordingly. Thus, in principle at least, there seems to be no difference between theoretical ethics and applied ethics on the score of dealing with real-life problems. And, as I suggested earlier, theory begins in the first place by attempting to provide an account of the "given" in moral experience. So theorists would say, I suspect, that real moral problems are important as a testing ground for ethical theory and that, in fact, theorists are constantly checking the implications of theory for actual moral judgments.

Having been alerted to these three caveats, we will proceed to an account of applied ethics' contribution to ethical theory. But now we are warned: I am not making any claims about what *in fact* theorists have learned from applied ethics, and I do not want to say that these contributions could have come *only* through applied ethics. Rather, I will simply be describing what I have found about ethical theory by doing medical ethics. In a word, it amounts to the *inadequacies* of theory that were not seen quite so clearly until theory was confronted by a plethora of real-life problems. Revealing an inadequacy is a contribution of sorts. Albeit somewhat negative, it is the contribution of putting a theory to the test.

Contributions

One contribution that should be mentioned, but not dwelt on, is that applied ethics has stirred up enormous popular interest in ethics. What is debatable, I suppose, is whether that constitutes a contribution. Philosophers tend to be suspicious of popularity; that may partly account for the disdain some have for applied ethics. In any case, philosophers of ethics no longer speak only in esoterica to each other, refining and honing the fine points of their theories independently of any contact with the world of real moral problems.

The next point, in one sense, is my only point about the contribution of applied ethics to ethical theory. I think it underlies everything I have to say about the matter. All the rest will simply be species of this generic point. The basic point is this: What applied ethics has to offer are real, pressing moral problems in all their fullness and relentlessness.

To be sure, as has been mentioned, theorists do use *realistic* cases to illustrate points, and they even test various claims against *realistic* examples. But these cases and examples are "realistic" not "real." They are usually carefully selected and highly abstracted. That is, the variables—as in any controlled experiment—are reduced to as few as necessary to make the point. Surely most of them involve either promising to return a book to the library or whether to save a drowning person when it will make one late for an appointment! Pretty thin stuff when compared to what daily confronts applied ethics!

Theory seems to survive by the grace of *ceteris paribus*. It is theory's way of holding back the real world's flood of relevant variables as it carries out an in vitro thought experiment on one or two selected variables—everything else, of course, "being equal." But in the real world, it seems, everything else is never equal. Indeed, entering the real world armed with ethical theory, one has an overwhelming experience of *moral gridlock*. There are relevant considerations and claims coming from every direction. Action seems at a standstill; theory seems immobilized.

That is the basic point, and what follows is really only illustration. It would be more interesting to claim that applied ethics can show theories to be inadequate or wrong. I do believe that claim, but it is difficult to prove. What

makes it difficult is that ethical theory can be so amorphous
and so malleable that nothing clearly counts against it.
Fudge factors are rampant, created *ex nihilo* to fill, smooth
over, or prop up any faltering theory.

So for now, we must be content with simply illustrating
that basic point, namely, that ethical theories generally are
inadequate when they confront the world of real moral problems. Some, of course, are more so than others. As we have
seen, the idea of "*simply* applying a theory" is misleading, if
not ridiculous. At best, "application" is the first step; it is
almost never enough for a problematic case. Besides all the
necessary conceptual work I have described variously as
"ground preparation" and "below the moral rules," much
must be filled in that is more properly the domain of theory.
I will mention five points.

1. One of the first realizations one has in applied ethics
is that there is seldom, if ever, a unique moral solution to a
problem. (Utilitarians and John Rawls would have us
believe there are unique solutions.) In real situations, there
usually are a number of equally moral alternatives. People
weight factors differently. What is a great evil to one is less
of an evil to another. Goods are weighted differently, so
different balances are struck. Similarly, persons have different priorities. Theories, though purporting to lead to a
unique solution, do not give sufficient determinates to do so,
and there is reason to believe that they *cannot* give sufficient
determinates. Among other things, that would require an
objective ranking of all goods and evils and the establishment of priorities among them. I suspect it is through being
immersed with real people in real decisions in a tight net-
work of claims and counterclaims from all directions that we
come to recognize forcefully that this has not and probably
cannot be done. As a procedural matter perhaps we should
continue to *search* for unique solutions, but that is not itself

without dangers. Fudge factors and even morally irrelevant factors can be aggressively manipulated to force unique solutions.

2. In doing applied ethics, one fairly quickly sees that some ethical theories or aspects of those theories are too simplistic to be of any help. They will approve of anything— as well as its opposite. Utilitarianism, for example, aside from all its other problems, impresses one as something of a free-for-all, finding goods and pleasures in the most fantastic places, with intensities and longevities and fruitfulness for further goods and pleasures that boggle the mind. Pick a number—any number—and have the greatest good come out to be whatever one wants. Like any theory should, it tells one what is morally relevant, but without precision, procedure, or control.

Even when a relatively more specific direction is derived from theory, it can collapse in the face of actual situations. For example, consider Mill's admonition that a man's liberty can be restricted only when his action might harm others. In the abstract that is powerful. But in the real world, the imprecision of the word "harm" makes the principle unusable. A person could be locked up for refusing a haircut, swearing in public, or wearing an ugly tie. Now, of course, we could begin to work on the concept of harm—drawing lines, making distinctions, digging deeper—and that is exactly the kind of thing we do in applied ethics. That's an example of what I have called "ground preparation." How one explicates "harm" will make all the difference in what actions get labeled moral and immoral. Theory is silent on the matter; it is applied ethics that sees the need, explicates the term "harm," and argues why some kinds of harm justify loss of liberty whereas others do not. (Of course, the argument might ultimately be based on utilitarian considerations.)

Theories overly simplistic or too general are quickly put to the test by real and complicated situations. A theory that proclaims freedom as the essence of morality, such that infringements on freedom are identical with immoral behavior, would soon be found out. As one tries to use such a system in making real moral decisions, one discovers—as in our previous example—that anything goes. Causing pain to a person is clearly immoral, but how is it an instance of limiting the person's freedom? Cheating is immoral, but does it limit the victim's freedom? If "infringing on freedom" can be conceptually stretched so as to include cheating and causing pain, then even giving a gift could be seen as "infringing on freedom" and hence immoral. And that is absurd.

3. Another common inadequacy, magnified in practical circumstances, concerns the scope of theory. To whom—or to what—does the morality pertain? A theory is like a formula, ready to function, awaiting only the assignment of the value of the variables, but we are not sure over what domain the variables range. Are fetuses included? Are animals? Are plants? Are corporations? Are all born humans? Are the senile? The comatose? Only rational beings? Then what constitutes "rational"? Are they competent? What constitutes "competence"? These are, once again, good examples of the concerns of applied ethics and examples of "ground preparation." These are conceptually complicated matters, and in explicating them we are in effect determining which actions will be considered moral and immoral. Questions of scope in applied ethics come as close as any to forcing theory to reconsider foundations.

Another "scope" issue might be locating the recipients of action. Effects of actions resound, resonate and ricochet throughout a society. For how much is the agent responsible? How much must he consider in his deliberation? Theory conveys none of the richness of these enormous complications in determining the effects of actions.

4. There is still another kind of inadequacy that comes to light under the relentless persistence of real situations. It is a matter of experience pushing theory toward greater refinement. This seems to me the ideal relationship of theory to experience. Theory grows out of experience, and then, transformed, returns to experience for a trial run. Shortcomings are found, and then it's back to the drawing board for further refinement.

An example is the growing awareness, through experience, that theory does not pay enough attention to the network of relationships into which we are woven in society. These relationships create subtle responsibilities, expectations, and obligations that are not factored in by the formulas of moral theory. The utilitarian calculus would probably come out the same no matter which child of several one decided to save. But a real-life situation would reveal the felt obligation to save one's own child. On the other hand, one has an obligation to break a moral rule with respect to his own child (say, causing pain, restricting freedom, depriving of pleasure) in a way he has toward no other person, namely, for punishment. Similarly, a representative government can and must break certain moral rules with respect to its citizens that citizens may not break with respect to each other.

The point is that theories, in effect, are not unlike formulas prescribing the interaction of equivalent, identical units. This is analogous to scientific theories conceiving of space—time as composed of units or points, completely identical with each other. But in this case those basic units are moral agents, and the problem is that ethical theories are not sensitive to the implications and understandings that go along with that "unit's" role or position in a complicated net-work of relationships. Working through the details of real situations makes one aware of the moral relevance of these roles and relationships. For example, in universalizing the maxim on which one acts, or on which one is basing an exception, specific roles and relationships may indeed be morally relevant and as such should be included in the description of the maxim on which one is acting. That is the sort of awareness that can lead eventually to the refinement of theory.

5. Another area brought to light by the press of real problems is the structure and components of moral disputes. One discovers that disputes are seldom disagreements over moral principles or rules, but are usually over facts, empiri-cal matters, or their interpretations. These are usually com-plicated by being stated in terms of probabilities, so that the debate is over whether a course of action is justified when there is such and such a benefit accompanied by such and such a risk. Perhaps ethical theory could benefit, again, by realizing how inadequate it has been in understanding and helping with the arguments of applied ethics. Theory could conceivably gain insights into the structure of moral argu-ment, into the determination and the balancing of goods and evils, and into the role of probabilities and conditions of uncertainty by examining the elements that comprise so many of the moral problems we confront in real life.

Quick Summary

1. On the surface there is a chasm between ethical the-ory and applied ethics. Each seems to have a life of its own: its own points, its own concerns, its own strategies, and its own concepts.

2. On closer inspection, a bridge-over-the-chasm can be found in the continuum that distinguishes components

within "ethical theory": metaethics, foundations, and systems (principles, rules, and bridging rules). It is this last component (i.e., working rules) that applied ethics particularly uses when it uses theory. But, nevertheless, the focus of applied ethics is on facts and the conceptual analysis relevant to adapting and refining moral rules for the particular subject matter at hand.

3. The contribution of applied ethics to ethical theory is not great, but it is important. The complexity and relentlessness of real moral problems show the weak points and inadequacies of theory, forcing it to more detail and specificity. Of course, if my continuum view of theory extends even to what I have called "ground preparation," then applied ethics makes many important and insightful contributions to theory. But if the continuum is stretched that far, then applied ethics is really a part of ethical theory, and the two are distinguished only by being at opposite ends of a spectrum.

Applied Ethics
Don't Change the Subject

Howard Brody

Introduction

I wish, in this paper, to explore the idea of applied ethics in medicine as a special type of ongoing conversation and to see how far this "conversation" metaphor can be pushed before it loses its utility. In listening to the preceding papers, I was struck by a number of observations that seem compatible with what I wish to do here. Accordingly, I shall begin by reviewing these observations briefly, to set the stage for what follows.

First: Several of my colleagues have, in one way or another, characterized applied ethics as: "First you get clear on the facts, and then you...." The comfortable assumption is that it's what comes after the ellipsis that is both interesting and "philosophical." But several speakers have shown, instead, that it is precisely "getting clear on the facts" that is both most important and least studied. If there is a method for "getting clear on the facts," what might it be like?

Second: Tom Murray, some time ago, offered a quick-and-dirty refutation of hospital ethics committees that goes like this: Either there is a principled way to derive an ethical resolution to a problem, or there is not. If there is, any single individual who knows the relevant facts and the principles in question can derive the answer. If there is not, no hospital

ethics committee could have a clear idea of what it is supposed to do. Therefore, hospital ethics committees are either
unnecessary or ineffective.

That a thoughtful philosopher of medicine would offer
such an observation shows, to me, two things. First, Art
Caplan's "engineering model" is *not* a straw man. Second, we
can just see how radical is Terry Ackerman's point that
perhaps a justified correct answer to a moral problem inherently requires a social interaction of some sort.

Third: Terry Ackerman also pointed out that a moral
community is defined by a commitment to shared norms of
behavior *within the process of inquiry.* If the members of a
hospital ethics committee are disagreeing violently about
whether it is permissible to stop tube feedings in an anen-
cephalic infant, for example, it is easy to discern the moral
principles that are *not* shared and which account for the
disagreement—commitment to biological life vs commitment to personal mental capacities, for instance. But what
are the deeper norms of behavior that *must* be shared in
order for the discussants to be *having the discussion* at all?

Fourth: Both Dan Clouser and Ruth Macklin reminded
us that the most fundamental role of ethical theory in
applied contexts may *not* be to point to an answer once the
facts are known. Instead, theories may play the role of
telling us which facts are indeed *morally relevant.*

Fifth: Dan Clouser, in his teaching of applied ethics,
typically reserves discussion of philosophical theory until
the end of the course. Coming at that time, theories serve to
summarize and organize moral considerations that the students had intuitively appealed to earlier in the course. This
leaves open the possibility that even discussants unschooled
in theory will nevertheless lay on the table, for considera-

tion, precisely the same factors as those with a strong theoretical background.

Sixth: Clouser further noted that an inadequacy of most philosophical theories, as currently accepted, is the way they tend to neglect moral duties arising from special relationships or role responsibilities. All of us agree that these duties are morally relevant considerations, despite the short shrift they receive from accepted theory. So by what means do they emerge as moral considerations in our deliberations, and how do we then deal with them?

With these six points in mind, I turn to a discussion of how hospital ethics committees might function.

Two Committees

Imagine the newly constituted ethics committees of two hospitals, Man's Greatest Hospital and St. Elsewhere's Hospital. (For purposes of discussing applied ethics in medicine, I prefer to use the hospital ethics committee as a starting point. We could as easily look at case consultations provided by hospital based ethicists, or continuing medical education programs conducted for hospital staffs in medical ethics; but the hospital ethics committee, as a relatively new and untried mechanism, promises to be a particularly interesting test case for issues in applied ethics in medicine.)

These two ethics committees have a great deal in common. Both have been established in part to review specific cases that arise in the hospitals. Both are designed to be broadly based, multidisciplinary committees representing most of the specialties and points of view present within the hospital staff. Furthermore, both have memberships dedicated to reading important background material in medical ethics and keeping up on current developments. They have studied the President's Commission Reports and are regular subscribers to the *Hastings Center Report, Medical Ethics Advisor*, and other pertinent periodicals.

In one important way, the two committees have decided to go in different directions. The committee at Man's Greatest Hospital is pursuing a thorough grounding in ethical theory. Philosophers trained in ethics have been brought in to conduct workshops, seminars, and other forms of training to impart the same expertise in ethical theory that is commonly . taught in philosophy classrooms. As a result, committee members aim to be adept at the vocabulary of utilitarianism, deontological ethics, natural law theory, Rawlsian theories of justice, and, indeed, all aspects of ethical theory that may be relevant to medical problems.

The ethics committee at St. Elsewhere's Hospital has taken a different tack. They have decided to forgo any in-depth training in ethical theory and consultation with professional philosophers, other than what might be suggested by reading specific articles in the journals and the reference materials mentioned above. Instead they have adopted a simple rule of procedure. They have agreed among themselves that they *will not change the subject.* That is, they will dedicate themselves to making sure that every point of view and every consideration that could conceivably be relevant to a particular case is put on the table, in the depth necessary for each member of the committee to understand its applications to the case at hand. They will continue to search for relevant considerations until they are convinced among themselves, after a thorough and rigorous investigation, that they have in fact included all relevant considerations in their conversation. Only then will they try to reach closure.

Furthermore, they will be willing to reopen their discussion of any case or any topic area (such as do not resuscitate orders, withholding life sustaining treatment, and so on) whenever a member of the committee feels that a relevant consideration or an important factual consideration not previously addressed has just been identified. They will also be willing to reopen conversation when a member of the committee gives reasons to argue that an inadequate priority judgment was made among considerations reviewed previously, or that important weaknesses or strengths in considerations were inadequately attended to.

They justify this procedural rule by claiming that, in practice, wrong decisions are reached most commonly because an incomplete set of factual data or moral considerations has been put on the table for discussion. They feel confident that they will be able to reach more reasonable conclusions by following their "don't change the subject" rule. Furthermore, they feel confident that, from the considerations laid out, they will be able to recognize the more relevant and more compelling ones even without a deep background in ethical theory. At least they feel this will be the case if they spend long enough investigating the considerations under discussion.

We can now ask the question: Which committee is likely to make better choices and why? By "correct choice" I will mean the choice that best coheres with the definition of ethics offered by Martin Benjamin and Joy Curtis; they claim that ethics is the attempt to answer the general question: "What, *all things considered*, ought to be done in a given situation?"[1] Even if this is not a satisfactory definition of ethics generally, it certainly ought to serve as a reasonable definition of applied ethics.

Applied Ethics as Conversation

In assessing the approach of the St. Elsewhere's Hospital committee, we might note that the primary conception of applied ethics, as the committee conceives it, is having and rigorously sustaining a particular sort of conversation. Initially, this conception may seem foreign to a view of ethics that has focused on the application of theories, the elaboration of rules and principles, and the making of specific decisions. However, the conversation metaphor has recently cropped up in interesting ways both in philosophical theory and in medical ethics itself, and so is worth a longer look.

The theoretical implications of "conversation" have been stressed in works by Richard Rorty[2] and Richard Bernstein.[3] They use this metaphor to do two things primarily—first, to demystify philosophical expertise and picture the philosopher more as an intellectual utility infielder rather than as someone who has access to the foundations of knowledge through indubitable epistemological theory; and second, to relate philosophy to the ongoing evolution of a particular culture, simultaneously denying that philosophy can hope to achieve an eternal and culturally neutral reference point from which it can have privileged access to truth. Since the works of the great philosophers of the past represent both insightful and trenchant commentaries on the intellectual development of the culture and major contributions to the further development of the culture, philosophers who have carefully studied and thought about these texts do indeed have a kind of expertise that they can offer to contemporary debates on applied issues. Haavi Morreim has noted that applied ethics is particularly philosophical because the rightness of an answer depends less on its content and more on the quality of the reasoning that leads up to it. Because philosophers have carefully attended in a professional way to the study of what constitutes good reasoning, they have special expertise to add to this enterprise.[4] But this sort of expertise is more modest than, and must be carefully distinguished from, a presumed expertise based on a theory that purports to identify the necessary foundations of knowledge in either science or ethics.

The metaphor of conversation has also been found useful within medical ethics. Jay Katz, in his book *The Silent World of Doctor and Patient,*[5] tries to replace a legalistic or adversarial concept of the process of informed consent with the more benign and humane picture of an ongoing conversation that defines the relationship between the physician and the patient over time. For Katz, as well, the metaphor of conversation serves several purposes. It regards the physician–patient relationship as ongoing and open-ended in important ways. It sees the patient's involvement in the process as participatory, rather than suggesting that the patient is a passive recipient of information. And it also attempts to demystify and delegalize the general notion of informed consent.

My exploration of the relative contributions of the conversation metaphor, as opposed to ethical theory as traditionally conceived, in the work of a hospital ethics committee thus can be viewed as an exercise in taking a promising concept, pushing it as far as possible, and seeing where its limits lie.

The Role of Theory

The approach taken by the St. Elsewhere's Hospital ethics committee does not necessarily imply that there is no useful role for theory in their deliberations. It does suggest, however, that theory will have a supportive rather than a central role in the conversations that comprise their applied

ethics activities. The particular conversations that constitute this committee's deliberations about particular cases or
topics will take place against the backdrop of the much
broader conversation that is the history of our western culture over the last several hundred years. It is (Rorty and
Bernstein argue) out of this larger conversation that theories of ethics have evolved. In the larger cultural conversation, certain recurring themes have emerged that identify
generally applicable characteristics of particular deliberations tending to identify certain actions as right or wrong.
Theories like "treat others as ends and never as means only"
and "seek to maximize the good for the greatest number" are
useful, pithy summations of decades or centuries of scrutiny
of particular actions, attitudes, motives, and consequences
of human behavior. Lengthy and vigorous attempts by
philosophers to establish the clear superiority of one of these
well-established ethical theories over all competing theories
have simply failed. Thus the question, "What ought to be
done in a given situation, all things considered?" instead of
"What ought to be done in a given situation, according to
theory X?"

These assumptions about ethical theory contrast mark-
edly with the position taken by Engelhardt in his recent
volume, *The Foundations of Bioethics*.[6] Rorty might be
tempted to say that Engelhardt has not yet awakened from
his Kantian slumber. Kant believed both that he could identify necessary preconditions for knowledge in ethics, and
that a thorough analysis of the rational will could yield a
detailed theory of the moral good of all persons. Engelhardt,
recognizing that we live in a pluralistic, secular society, has
abandoned the hope that there could be any universalized,
rationally based theory of the good of an individual human
life. He admits that this will be a culturally and socially
relative phenomenon. However, he adheres to Kant's faith

in establishing the logical priority of certain principles that provide the necessary foundations for any possible knowledge of ethics. Despite the fact that these principles either are so general as to be largely uninformative or else are quite controversial in their application, Engelhardt holds that the investigation of the foundations of ethical theory has priority over the determination of particular goods in particular social circumstances. From his two-tiered conception of ethics (in which generally universalizable and foundational ethical theory yields general side constraints, whereas particular social and cultural circumstances dictate a particular individual's ideal of the good life), our two hospital ethics committees might reasonably expect to get little help.

An approach to ethical theory more congenial to the work of our two committees is found in the commonly used textbook by Beauchamp and Childress.[7] These authors note at the outset that one of them is a secular utilitarian and the other is a Christian deontologist, insofar as their theoretical allegiances are concerned. Despite this difference, they argue that they can agree upon some basic principles that can, in turn, inform useful practical moral rules for medical ethics. If one claims that the conversation of Western culture has been deeply informed by Christian theology, but also of late has been significantly influenced by utilitarianism and related political theories, then the coherence of these two author's views at the level of moral principles and moral rules is hardly surprising. The lesson for the ethics committees is that theoretical differences need not preclude agreement on individual cases.

More importantly, we begin to see some justification for the apparent disregard of formal training and ethical theory on the part of St. Elsewhere's Hospital ethics committee. They might assume, based on the fact that ethical theory reflects important strands and themes that have shaped the broader conversation of our culture, that any reasonably informed and educated person in that culture will already have a fairly developed sense of at least some of the moral factors and basic considerations that make up different ethical theories. True, it is unlikely that any one of them has a comprehensive overview of these different variables, assuming that he or she has not made a detailed study of philosophical ethical theory. But here again they have faith in the diversity of views and backgrounds represented on a multidisciplinary committee. It is highly unlikely that an extended conversation among them will not eventually produce the same range of considerations that would have been put on the table had a skilled philosopher identified all the important theoretical strands of the topic at hand. For example, in dealing with a specific case, it is highly unlikely that all of them will restrain themselves to talking about benefits, and that no one will even raise the issue of individual rights. It is highly unlikely that all of them will restrict themselves to considering the avoidance of harm, and that no one will ever raise questions about fairness and the need to treat like cases similarly.

Rules of Conversation

It is at this point that the operational rule, "don't change the subject," becomes clearer in its importance. This rule might be said primarily to distinguish the peculiar sort of conversation that makes up ethical deliberation from other sorts of conversations that we might label bull sessions, social chitchat, or exchanges of prejudices and opinions without movement toward resolution or understanding. In these other forms of conversations, people change the subject as soon as they begin to get uncomfortable with the need to explain or justify their beliefs, or as soon as a particular topic has been under discussion for a period of time. It is precisely when changing the subject occurs, according to the St. Elsewhere's committee, that one is likely to make a wrong choice. One cannot be assured that all relevant moral considerations have been placed on the table for discussion before the subject was changed; or even if all considerations were laid out, that they have been critically dissected to see which are worthy of great weight and which represent emotional biases that ought to be put to one side. This committee has reasoned that if they can assure themselves that they will not change the subject until they are sure that they have exhausted everything they can think of to say about a particular course of action and all the considerations that may be relevant to it, they will have considerable confidence that they will reach an appropriate and defensible position.

We can now begin to assess the differences that will occur in the deliberations of the two hospital committees. It may very well be the case that the conversation at Man's Greatest Hospital will be more productive and more efficient. Since ethical theory is a succinct summary of how many similar conversations have gone throughout the history of our culture, knowledge of ethical theory may be of great assistance in streamlining and enhancing the depth of the conversation that occurs in that committee. Alternatively, the Man's Greatest Hospital committee may occasionally be misled. Because any single theory is an incomplete survey of all the factors that may be morally relevant in a particular case, overly relying upon a particular pet theory may shut off conversation before other competing considerations have been given attention. In particular, if a case presents unique features such that existing theories and maxims are of little direct use, these individual aspects of the case may be given short shrift in the discussion of the Man's Greatest Hospital committee.

Unfortunately, we cannot rule out the possibility that the St. Elsewhere's committee may end up making idiosyncratic decisions that systematically ignore certain basic moral considerations, simply because the limited experience and backgrounds of the members of that committee, or peculiar features of that institution, are such that relevant considerations are simply not thought of when the committee meets to deliberate. It might be the case, for example, that the various members of the committee, when they think of possible harms to be suffered by individual patients, think solely in terms of physical and emotional harm and do not consider affronts to basic human dignity such as violations of the right of privacy. Over time, this blind spot could become self-perpetuating. There can be no doubt that an appropriate injection of ethical theory could remove this moral blind spot. According to our metaphor, however, the main question is whether a broader conversation will occur among many hospital ethics committees that will allow them to compare notes and to see the bases upon which they have reached their respective decisions in individual cases. If the St. Elsewhere's committee, in addition to carrying out their own internal conversation, eventually participate in a broader conversation of this type, the error will speedily be exposed.

Finally, according to our metaphor, expertise in medical ethics does not consist in logical skills of argumentation or having access to a particular theory that provides foundational knowledge of ethics. Instead it consists of having attended carefully to the broadest and most inclusive conversations in the area of medical ethics over a reasonable period of time. As indirect support for this, I suggest that we consider the role that certain "war-horse" cases have played in our understanding of medical ethics and in our off-the-cuff judgments of the expertise of our colleagues. Imagine that at a meeting like this we get ourselves into a conversation with a teacher of ethics whom we have not met previously. This person appears to be well schooled in theory. When we make reference to "the Galveston Burn Victim Case" or "the Johns Hopkins Case," however, he indicates a total ignorance of what we are talking about. We are likely to conclude that this person is a rank newcomer and has limited expertise in medical ethics. What we are saying is that this person does not seem to have participated in the same sorts of conversations that most of us have had over the last 10 or 15 years.

Methodological Reflections

To this point I have been trying to use the conversation metaphor as a deliberately atheoretical way to demystify some issues in applied ethics. I would now like to return to more theoretical language to borrow additional support for the potential utility of the metaphor.

One analysis of applied ethics in medicine that seems most congenial to what I have described as a conversation model is Jonsen's contention that clinical ethics as now practiced is a form of casuistry, rather than the application of ethical theory, at least as ethical theory has come to be envisioned in the 19th and 20th centuries.[8] In addition to describing the nature of casuistry, Jonsen has added several observations that seem to me to cement its relationship to the metaphor of conversation.

First, the casuists, during the heyday of their enterprise, produced a voluminous literature but extremely little explicit discussion of methods. In effect, they assumed their methods would be clear to anyone who carefully read their analyses of cases. Jonsen notes in an illuminating sentence, "Indeed they may not have thought much about method, as we understand the idea."[9]

Second, casuistry was rooted less in philosophy and more in the traditional discipline of rhetoric. The grounding in rhetoric, in turn, led to the concept of "probable certitude" that was commonly used by casuists. That is, philosophical methods were seen as most appropriate when an answer could be known with certainty; whereas rhetoric was seen as the discipline of choice when a conclusion could be reached not with certainty but, because of the convolutions of practical affairs, only with reasonably high probability. Despite not being rooted in philosophy as a basic discipline, however, the casuists borrowed freely from whatever philosophical theories of ethics and morals they could get their hands on. In an eclectic fashion, they borrowed equally from theology and legal sources. For them, anything that promised to increase their "skill of interpreting the case"[10] was fair game.

Third, casuistry, as practiced during its days of greatest legitimacy, was actively and persistently self-correcting. The casuists would give critical commentaries on the case analyses of their fellow casuists; and the others, as soon as they read these critical analyses, would frequently go back and revise their own views on a particular case or set of cases.

These aspects of casuistry lead Jonsen to recommend that modern clinical ethics can learn a lot from the casuistry model. In particular he highlights the following features: (1) reliance on paradigm cases; (2) reliance on broad consensus to reach agreement; and (3) the desire to reach "probable certitude." (Following my colleague Martin Benjamin, I prefer the term "provisional commitment" as against "probable certitude," because the former appears less self-contradictory and also indicates that one must take full moral responsibility for decisions even when the certainty of the knowledge base cannot be presupposed.)

An interesting implication of Jonsen's account of casuistry is that the "skill of interpreting the case" of which he speaks may be more a pattern-recognition skill than a logical-deductive skill. In today's jargon, we may be talking more about right-brain and less about left-brain functioning. I find personally that when I try to envision what Jonsen is saying about the casuistical method, I start thinking in terms of geography, mapping, and putting cases side-by-side in an orderly array within a two-dimensional space. Though this counts as a crude hunch rather than careful reflection on the nature of casuistry, further elaboration may help to define the differences between the so-called casuistical method and the method of deducing implications from theories.

One important corrective needs to be added to the picture of casuistry in order to make clinical medical ethics coherent within a modern, secular, pluralist world view. Jonsen notes that, according to the scholasticism of their time, the original casuists were basically natural law theorists. For them, natural law could give rise to maxims, which then could be used to organize cases, depending on how clearly or how poorly they reflected the moral maxims in practical application. Because natural law was immutable, however, one could not reason backwards and alter the maxims based on what one learned from specific cases. Today, the only approach that seems to make sense here is something like John Rawls' concept of wide reflective equilibrium.[11] The method of wide reflective equilibrium insists that we must reason in two directions, both from the basic principles to revised considered judgments and from important considered judgments to revised basic principles. Furthermore, both must be in some sort of creative tension with the best available psychological theories of human nature. To meet the criticisms of those like MacIntyre,[12] who argue that a system of ethics cannot be coherent unless it includes some discussion of an ultimate end or *telos*, wide reflective equilibrium would also include as its *telos* the ideal of a just, well-ordered society.

Further indirect support for the conversation metaphor is found in some reflections on clinical ethics by Morreim.[13] For example, a possible objection is that conversation will inevitably fail to clarify issues because the participants won't be properly skilled in distinguishing moral from non-moral elements. A good deal of the early writing on medical ethics in the 1970s was devoted to distinguishing moral or values issues, where patients should have the expertise to make their own decisions, from technical issues, where physicians naturally had more authority. Unfortunately, Morreim comments, from her perspective as a hospital-based philosopher, it's getting harder and harder to distinguish the moral from the nonmoral elements in medical decisions. The same conclusion was reached in an analysis of common objections to medical ethics among primary care physicians.[14] As noted above, Morreim's conclusion is that clinical ethics is truly philosophical because it depends on good reasoning. She feels that good reasoning consists primarily of three elements—a proper understanding of the situation and issues; detailed identification of all important factors and options; and logical precision, which consists primarily of noting the strengths and weaknesses of various positions. The important lesson I draw from this is that ethical theory, as such, may aid somewhat in the first and third, but it is not specifically a part of good reasoning as defined by Morreim. Indeed, as already noted, overreliance on theory may hamper the second element.

Conclusion

Although I indicated earlier that the limits as well as the possibilities of the conversation metaphor must be explored, space precludes a detailed analysis of limits, so a discussion of these will have to await another occasion. It may be that unless the type of explicit methodology that comes from a firm grounding in ethical theory can somehow be inserted, it will remain a hopeless task to distinguish a rigorous and disciplined moral conversation from other unsatisfactory forms of discourse. In addition, it may be that a rule such as "don't change the subject" will work satisfactorily only when one is dealing with a particularly sophisticated committee or community, and the value of such a rule will be lost on a less sophisticated group, which may be closer to the norm as different hospitals across the country begin to try these new approaches. In this regard, however, the comments of other philosophers in clinical settings are heartening, since they seem to agree that moral conversation can be both philosophical and disciplined without necessarily relying on the application of ethical theory.

References

[1]M. Benjamin and J. Curtis (1981) *Ethics in Nursing,* Oxford University Press, New York, NY, p. 9.

[2]R. Rorty (1979) *Philosophy and the Mirror of Nature,* Princeton University Press, Princeton, NJ and (1982) *Consequences of Pragmatism,* University of Minnesota Press, Minneapolis, MN.

[3]R. J. Bernstein (1980) "Philosophy in the Conversation of Mankind," *Review of Metaphysics* **33,** 745–775.

⁴E. H. Morreim (1986) "Philosophy Lessons from the Clinical Setting: Seven Sayings that Used to Annoy Me," *Theoretical Medicine* **7**, 47–64.

⁵J. Katz (1984) *The Silent World of Doctor and Patient,* Free Press, New York, NY.

⁶H. T. Engelhardt (1986) *The Foundations of Bioethics,* Oxford University Press, New York, NY.

⁷T. L. Beauchamp and J. F. Childress (1983) *Principles of Biomedical Ethics,* 2nd ed., Oxford University Press, New York, NY.

⁸A. R. Jonsen (1986) "Casuistry and Clinical Ethics," *Theoretical Medicine* **7**, 65–74.

⁹*Ibid.,* p. 68.

¹⁰*Ibid.*

¹¹J. Rawls (1980) "Kantian Constructivism in Moral Theory," *Journal of Philosophy* **77**, 515–572. *See also,* N. Daniels (1979) "Wide Reflective Equilibrium and Theory Acceptance in Ethics," *Journal of Philosophy* **76**, 256–282.

¹²A. MacIntyre (1981) *After Virtue,* Notre Dame University Press, South Bend, IN.

¹³Morreim, *supra,* note 4.

¹⁴H. Brody and T. Tomlinson (1986) "Ethics in Primary Care: Setting Aside Common Misunderstandings," *Primary Care* **13**, 225–240.

Philosophical Ethics and Practical Ethics

Never the Twain Shall Meet

Barry Hoffmaster

A former dean of one of Ontario's medical schools recently described biomedical ethics as the biggest growth area in medicine. Regardless of the accuracy of that assessment, the rapid, exponential expansion of the field cannot be denied. Reflection concerning the nature of the field unfortunately has not kept pace with activity in the field. Only lately have questions about what biomedical ethics is and how biomedical ethics ought to be conducted received serious attention. At the same time, answers to these questions have begun to polarize.[1] At one pole are advocates of philosophical ethics, those who see biomedical ethics as essentially the application of standard philosophical positions, utilitarianism or Kantianism, or principles derived from standard philosophical positions to the facts of cases. Ramsey, for example, remarks:

> ...serious ethical inquiry and discourse begin only when we discover we are in disagreement about what we ought to do. Then we are forced back upon our premises, and we must seek...to find agreement at a deeper level. We must ask about what makes anything right. We need to find out if we can agree upon the right-making or wrong-making features of

moral attitudes, actions, roles, or relations, before returning
to the specific case where we first disagreed....That is ethics
proper.[2]

Biomedical ethics, in this view, is applied moral philosophy.[3]

At the other pole are those who reject the pertinence of
ethical theory to actual moral problems. The most striking
feature of real moral problems is that they need to be
resolved, but the gap between the particularity of cases and
the generality of philosophical theories and the abstract
principles they spawn precludes resolutions. Advocates of
practical ethics, therefore, appreciate the need to develop an
alternative approach to biomedical ethics, one that is concerned with cases rather than theories and that consequently is more useful in the resolution of moral problems.
As Ackerman observes, "bioethics must be helpful—helpful
in determining what to do under the demands of concrete
moral situations."[4]

This paper adopts the latter position. My critical claim
is that the problems with traditional moral theories and
their putative application to concrete problems are more
serious than mere gaps or inadequacies that could be reme-
died by ingenious philosophical tinkering. One needs, therefore, to produce an alternative conception of morality that
enables ethics to contribute to the resolution of moral problems without reducing it to blind intuitions, gut feelings, or
appeals to authority, and thereby allowing it to wallow in
inconsistency. My positive contribution is a suggestion as to
how that alternative can be developed.

The most siginificant contribution that the burgeoning
interest in biomedical ethics, business ethics, and ethics in
the professions generally could make to philosophy is a
deeper, more realistic appreciation of the nature of morality
and moral epistomology. Our understanding of morality has
been shackled and distorted by forcing it to conform to the

dictates of ethical theories and by approaching it in light of the philosophical dogmas that epistemology comes first and epistemology means foundationalism. The current concern with real moral problems presents philosophers with an opportunity to develop a fresh picture of morality and to point moral philosophy in a more productive direction.

Criticism of Philosophical Ethics

The assumption animating philosophical ethics is that moral knowledge is theoretical. As Noble has observed, 'Moral philosophers today assume that the possibility of genuine moral knowledge is the same as the possibility of valid normative ethical theories."[5] The program is familiar. The motivation of philosophers is to introduce objectivity, to the maximum extent possible, into morality. Objectivity presupposes rational justification, and rational justification presupposes foundationalism of one sort or another. At least four varieties of foundationalism can be distinguished (*see* Figure 1).[6]

The fundamental norm in A) could be act utilitarianism or the categorical imperative; in B) it could be rule utilitarianism. The decision procedure in C) and in D) could be some version of contractarianism or an ideal observer theory. The fundamental norms in these four alternatives are assigned three functions: 1) to produce decisions in A; 2) to justify rules or principles in B, C, and D; and 3) to provide comparative weights for principles when they conflict in C and D.

These schemata incorporate the trinity of philosophical ethics—autonomy, unity, and rationality. Morality, as incarnated in philosophical ethics, is independent of other theoretical disciplines as well as the social and cultural milieu in which it operates,[7] and this autonomy is explained in terms of its foundational norms: Morality is a distinctive theoretical enterprise because it rests on distinctive founda-

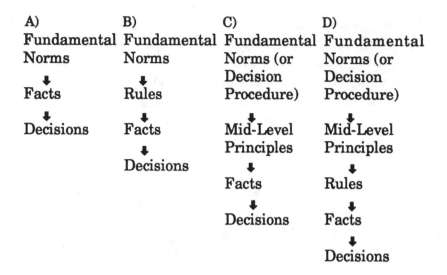

Fig. 1. Varieties of Foundationalism.

tional norms. At the same time, these norms explain the
unity of morality. Decisions, rules, and principles are moral
rather than legal or prudential because they can be traced
back to one and the same foundational norm through a chain
of reasoning analogous to the legal positivist's chain of
validity. The rationality of the enterprise is manifested in
the derivation of more specific rules and principles from
higher order, more general norms, and in the deductive
reasoning that subsumes facts under rules or principles to
yield decisions.

What is wrong with this conception of morality? There
are a number of counts in the indictment of philosophical
ethics. The only surprising fact about these counts is the
persistent indifference of philosophers to them.

Count 1. To begin, one would think the failure of the program would be more disconcerting. Despite the extensive
and extended discussion that philosophical moral theories
have received, there remains no uniformly accepted moral

theory. As in other areas of philosophy, every philosopher has his or her own favorite moral theory. More worrisome, however, is that, notwithstanding the initial appeal of Rawls' notion of reflective equilibrium, there is no accepted methodology for deciding between competing moral theories. Moreover, the attraction of reflective equilibrium itself, in my opinion, derives largely from the commodious role it assigns to specific moral judgments. It shifts the burden of justification from the rational defense of abstract principles to the plausibility of judgments about particular issues.

In addition, given the nature of philosophical ethics, any proposed method of interpersonal justification falls prey to the same problem that philosophers of science confront—the theory-ladenness of data. Hesse summarizes the general point nicely:

> ...the work of Wittgenstein, Quine, Kuhn, Feyerabend, and others has in various ways made it increasingly apparent that the descriptive language of observables is "theory-laden," that is to say, in every empirical assertion that can be used as a starting-point of scientific investigation and theory, we employ concepts that *interpret* the data in terms of some general view of the world or other, and this is true however apparently rooted in "ordinary language" the concepts are.[8]

If this is true of empirical assertions, it is *a fortiori* true of normative or evaluative assertions, and it poses as serious an epistemological problem in morality as it does in philosophy of science.

Count 2. A second difficulty is that the notion of rational justification is ambiguous. To say that a moral decision or judgment is rationally justified can be understood in a weak and a strong sense. In the weak sense it means only that the decision is not arbitrary, that is, that there is a reason for it. In the strong sense it means that the decision is correct. Golding, in his book on judicial decision making, says that judges give reasons for their decisions to insure that the decisions are not arbitrary, and these reasons are intended as justifications of the decisions.[9] But nonarbitrariness does not entail correctness. Feinberg likewise distinguishes between a decision being consistent with reason, that is, not irrational, and a decision being dictated by reason.[10] Again, a number of competing positions on a moral issue can be consistent with reason. Interpersonal objectivity requires that one position be dictated by reason, and philosophical ethics cannot produce rational justification in this strong sense.[11]

Someone might object, naturally, that philosophical ethics is not concerned simply with reasons for moral decisions but with *good* reasons for moral decisions. This amendment does not help, however, because the notion of a good reason is ambiguous. If a good reason is simply a nonarbitrary reason, interpersonal objectivity is missing. If a good reason is a conclusive reason, philosophical ethics has yet to establish the existence of this kind of reason.

Count 3. Philosophical ethics cannot account for the diversity of moral judgments. The standard explanation here is that the same moral principles produce different results in different societies or cultures because the facts of cases are different in those societies or cultures. But this presupposes an untenably sharp distinction between theory and fact. As well, the differences are more often evaluative than factual. It was reported recently, for example, that England refused to allow the 12-year-old wives of Middle Eastern university students to enter the country. There is no dispute about age or marital status, so the objection must be moral, involving some notion of harm to these 12-year-olds, or some account of intrinsic moral wrongness. Such cultural differences pose a dilemma for philosophical ethics. Philosophical ethics either must insist on its claim to intersubjective and intercultural objectivity, at the risk of being charged with moral imperialism, or it must resign this claim. That cultural differences are problematic for philosophical ethics is not surprising because philosophical ethics forces morality to be an atemporal, asocial, and acultural phenomenon.

Count 4. Recent feminist criticisms have called into question both the hegemony and the social/cultural autonomy of philosophical ethics.[12] Regardless of the plausibility of a claim about gender differences in morality, this work exposes the value-laden nature of philosophical ethics. Rationalistic morality is one kind of morality, not morality itself, no matter how many times the authority of Kant is invoked. A morality of caring and responsibility grounded in the primacy of relationships is another. Philosophical ethics is embarrassed by challenges to its implicit assumptions, but when the challenge arises from empirical evidence rather than *a priori* argument, the embarrassment is doubled.

Count 5. Insofar as moral principles are viewed as linguistic entities, they suffer from the same generality and vagueness that infect many legal standards. An eminent scholar of Canadian constitutional law recently remarked that the test for the exclusion of evidence seized in violation of a right guaranteed by the Charter of Rights and Freedoms, namely, that the evidence should be excluded if admitting it would "bring the administration of justice into disrepute," is "meaningless."[13] In certain respects, of course, this is an overstatement. But in other respects there is an important point here. Such abstruse terms acquire content through their application to concrete problems. As they stand, they are little more than formal placeholders. Whether and how they apply to a particular problem is a matter for decision, a decision that can be guided only in the weakest way by the terms themselves.

The same is true for the general principles frequently invoked in biomedical ethics, for example, principles of autonomy, beneficence, nonmaleficence, and justice.[14] Their application requires choices that cannot be guided by the principles themselves. Hart explains beautifully why the guidance provided by general standards is limited in the law, and his argument is equally applicable to morality:

> ...the necessity for...choice is thrust upon us because we are men, not gods. It is a feature of the human predicament...that we labour under two connected handicaps whenever we seek to regulate, unambiguously and in advance, some sphere of conduct by means of general standards to be used without futher...direction on particular occasions. The first handicap is our relative ignorance of fact: the second is our relative indeterminacy of aim. If the world in which we live were characterized only by a finite number of features, and these together with all the modes in which they could combine were known to us, then provision could be made in advance for every possibility. We could make rules, the application of which to particular cases never called for a further choice. Everything could be known, and for everything, since it could be known, something could be done and specified in advance by rule.[15]

In other words, when dealing with concrete problems, the help provided by philosophical ethics runs out precisely where and when it is needed.

Count 6. One of the most persistent arguments in favor of philosophical ethics is that, in the absence of moral theory, one has no way of determining what facts are morally relevant. This virtue of philosophical ethics is vastly overstated, however, because, as was claimed in Count 5, the guidance provided by a moral theory runs out precisely when it is most needed. Gottlieb explains why in the context of a legal system:

> Disputes and conflicts of interests are a social fact, and courts are called upon to settle not only those aspects of disputes which are noticed in the *protasis* [the introductory clause

expressing a condition] of legal rules. If courts were to disregard "significant" factual distinctions, the cases they would then be deciding would bear little resemblance to the issues agitating the parties. Legal decisions are designed to settle issues arising out of actual situations which are themselves the product of complex and concurrent valuations. The reduction of the material facts of a case to *only* those facts which correspond to the *protasis* of existing rules of law, must, therefore, contradict the very function and rationality of the institution of adjudication....[16]

Gottlieb generalizes this conclusion to any system of rules designed to regulate behavior: "...the application of rules belonging to a system which regulates...necessarily presupposes resort to concurrent standards of valuation as well...."[17] It is these supplementary, concurrent standards that are really doing the work, but both the work they do and the values they infuse into the process of moral decision making are beyond the scope of philosophical ethics.

Count 7. Philosophical ethics cannot account for the "deeper" or "hidden" values that function at a cultural level and figure prominently but inconspicuously in decisions. Empirical research has identified a number of value orientations that are not mentioned in philosophical or biomedical ethics.[18] For example, with respect to time, is one oriented primarily toward the future, the present, or the past? With respect to the relationship between individuals and nature, does one see people in harmony with nature, people controlling nature, or people subject to nature? Does one's way of life emphasize doing or being? With respect to the relationship between persons (an issue that is discussed in philosophical and biomedical ethics), is the individual, the group, or lineage and hierarchy foremost?

An example will illustrate how these deeper values function in decisions.[19] A 23-year-old native woman from a reserve in the Queen Charlotte Islands had her third child in Vancouver. The delivery was difficult. A mid-forceps delivery was required because a transverse arrest occurred. Her obstetrician recommended a postpartum tubal ligation, and she agreed but subsequently regretted the decision. She did not understand why she could not have any more children, not because of an intellectual deficit, but because the explanation that sterilization would be permanent simply did not register with her. The notion of artificially ending her child-bearing capacity was inconceivable to her because of the values she held.

Her doctor was oriented primarily toward the future and saw people, undoubtedly doctors in particular, as firmly in control of nature. Because a sense of individual achievement was important to him, he was concerned about what this woman could accomplish (or, more accurately, could not accomplish) with three children. He also was thinking of society and the cost to society of this woman's children. The woman, on the other hand, was oriented primarily toward the present and regarded people as subject to nature. She came from a matriarchal culture in which having babies is very important. In this culture individual accomplishments are not necessary; what is important is being a warm, loving mother.

Because the doctor was dealing with only the individual and her body, he thought he needed only her consent. The woman would have liked to have spoken with her mother and aunts, however, and would have followed their advice. When this woman returned to her reserve, the emotional reaction was mixed. The others were happy about her new baby, but sad that she could have no more children. Philosophical ethics ignores the kinds of deep-seated cultural values that separated this woman and her doctor. And even if it did try to account for them, it would be incapable of proving which values are "dictated" by reason.

Count 8. Philosophical ethics cannot explain the more problematic features of practical ethics, for instance, why a problem is perceived as a moral problem at all, and why different perspectives are adopted on moral problems. With respect to the former, what is it that triggers ethical concern and analysis? Why, for example, is in vitro fertilization hotly debated, but expensive microsurgery to reconstruct Fallopian tubes is ignored?[20] The best example of the latter is abortion. Is abortion a matter of protecting the life of the unborn, a matter of women controlling their sexuality, or a matter of the distribution of power in society?[21] Philosophical ethics assumes that moral problems come prepackaged and neatly labeled, waiting only to be subjected to technical analysis.

An Alternative Epistemology

The picture of morality that emerges from philosophical ethics is reminiscent of the view of law as a "science" that prevailed not so long ago and was attacked by legal realists. Law was seen as a discipline based on a limited number of principles that could be learned by reading court decisions and that could be used to find "logical" answers to questions. Law existed in its own sphere, free of outside influences, and legal reasoning operated solely within that narrow realm. Law was regarded as a self-contained, independent, rational system. Similarly, morality, as conceived by philosophical ethics, consists of a limited number of principles that can be learned by reading ethical theory and that can be used to find answers to moral problems. Philosophical ethics is even purer, though, because its reasoning is more *a priori*. Rather than studying the resolutions of cases and deriving prin-

ciples from cases, rationally grounded theories are con-
structed. But, as Putnam has noted, rampant *a priorism*
leads to extremism:

> Philosophers today are as fond as ever of apriori arguments
> with ethical conclusions. One reason such arguments are
> always unsatisfying is that they always prove too much; when
> a philosopher 'solves' an ethical problem for one, one feels as
> if one had asked for a subway token and been given a pas-
> senger ticket valid for the first interplanetary passenger-
> carrying ship instead.[22]

One of Putnam's examples of such philosophical extremism
is Nozick's libertarianism, which "contradicts the moral
outlook of the whole Western tradition" and ignores the
trade-off of interests that have been "central to our moral
practice."[23]

Some may believe that extremism in the service of
rationality is no vice. This extremism results, however,
because moral theories take one insight (for example, that
human dignity should be respected, or that the greatest good
for the greatest number should be promoted), elaborate and
defend it in sophisticated and ingenious ways, and then
claim that this exhausts morality. Morality is homogenized.
Not only is morality rendered substantively homogeneous, it
becomes epistemologically homogeneous. All moral questions have a correct answer, namely, the answer dictated by
the moral theory.

Morality need not be seen as epistemologically black or
white, though. There are at least nine positions, ranging
from extreme skepticism to extreme objectivism:

1. Wrong answers exist/can be identified for no moral
questions.
2. Wrong answers exist/can be identified for some moral
questions.

3. Wrong answers exist/can be identified for all moral questions.
4. Better and worse answers exist/can be identified for no moral questions.
5. Better and worse answers exist/can be identified for some moral questions.
6. Better and worse answers exist/can be identified for all moral questions.
7. Right answers exist/can be identified for no moral questions.
8. Right answers exist/can be identified for some moral questions.
9. Right answers exist/can be identified for all moral questions.

Philosophical ethics sees the epistemological choice as between 7 and 9. But if one accepts that the content of morality can be variegated, one also can accept epistemological pluralism in morality. For some areas or some kinds of disputes, there might be right answers. For others, there might be only better or worse answers. For still others, there might be only wrong answers. And for some, there might be no answers.

What is needed is a conception of morality that does not regard it as substantively and epistemologically homogenous. What is also needed is a conception of morality that infuses it with rationality but that stops short of *a priorism*. In both respects such a conception would be more in touch with our moral practices and traditions.

A conception of morality that meets these constraints has been provided recently by Kekes.[24] I offer this not as the account of the "true" morality, but as an illustration of the kind of thinking that needs to be done about morality. Kekes' account has two virtues: it provides a theoretical interpretation of morality that is sensitive to actual moral experience, and it allows one to make sense of some of the salient features of practical ethics. The starting point of Kekes' position is the truism that there must be a limit to chaos and discontinuity for society to exist. The prosaic dictates of everyday morality prevent rampant social disorder and conflict.

Kekes focuses on commonplace morality rather than on moral conflicts, and sees this morality as consisting largely of intuitive responses expressed in spontaneous judgments. These are not intuitions in the classical sense. They are not self-evident, certain, unconditional, or the product of a moral sense. Rather, they are the same kind of phenomena as appreciating mathematical truths or recognizing patterns, colors, sizes, sounds, and persons. The object intuited is a situation, that is, an interpretation of nonmoral facts, and that interpretation involves cognitive, emotive, and volitional elements.[25]

Intuitions are immediate; there is no conscious inference, reflection, or thought involved. Intuitions occur routinely, so they are not to be sought in cases of conflict. Intuitions are imperatives, so they are practical and action-guiding. Intuitions are presumptive; they establish a *prima facie* case for a situation being as it is intuited. Intuitions are interpretive; they involving seeing *as*, not just seeing. Intuitions are unquestionably accepted; one is not making an inference and not entertaining a hypothesis, so one is not in need of evidence beyond what one already has. Intuitions are fallible; the interpretation of a situation may be mistaken.

The crucial problem, of course, is how such intuitions can be rationally or critically assessed. How can one adduce reasons for or against an interpretation of nonmoral facts? The first point is that reasons are appropriate. The notion of "seeing as" is to be understood along the lines of seeing a color as blue rather than seeing a figure as a rabbit or a duck.[26] The latter is arbitrary, but the former involves subsuming an object under a concept, so it is a matter of interpretation not whim. The reasons behind an interpretation are derived from a shared moral tradition:

> The justification of particular moral intuitions is that they derive from a sound moral tradition and that the unquestioning acceptance of what strikes a moral agent as obvious is shared by fellow members of the tradition who are involved in or witness the relevant situation.[27]

The justification of a moral tradition, in turn, is that "it fosters conduct that leads to as favorable balance of benefit over harm for its members as possible given the context."[28] Moral traditions are justified, in other words, in terms of whether they provide external and internal goods. External goods are related to traditions as instrumental means are to ends. Customary, commonplace morality is justified if it is instrumental in obtaining such goods as the protection of life, security, and property, some freedom, and some good will for those subject to that morality. These goods are required for the flourishing of life in any society. Internal goods are related to traditions as constitutive means are to ends, or as parts are to wholes. These goods define a tradition from within and can be as diverse as the many forms that life can take. Racist, Stalinist, and criminal moral traditions are not justified because they fail to provide the internal and external goods required for good lives.[29]

Two kinds of moral mistakes are possible in Kekes' account. First, one could misjudge the significance of non-moral facts because of self-centeredness or selfishness. Second, one could confuse moral contexts and, for instance, rely on intuitions in a context in which they are inappropriate. The moral practices of society form a continuum. At the private, personal end of the continuum, the issue is self-direction, how one ought to lead one's own life. Conflicts at this end are primarily between different moral traditions and thus involve choices among different internal goods. At the public, impersonal end, the issue is decency, how relationships between strangers ought to be governed. This is primarily a matter of securing external goods. But decisions about insuring external goods are largely settled (so settled, in fact, that they often are backed up by criminal law), and it is these settled decisions that are reflected in moral intuitions. In between the issue is intimacy, how relationships between loved ones and relatives, for example, ought to be conducted. Moral subjectivism or pluralism is not spread evenly over this continuum. There is more room for it at the private, personal end than at the public, impersonal end. Different moral contexts need to be distinguished, therefore, because they differ both substantively and epistemologically.

Kekes offers no account of how moral conflicts are to be resolved. His view nevertheless has many attractive features. Its methodology makes sense. One is more likely to understand morality by focusing on the common and prosaic than on the rare and dramatic. He provides a persuasive justification for commonplace morality reminiscent of Hart's argument that any legal system must contain laws protecting persons, property, and promises.[30] His view captures the heterogeneity of morality by locating it in a social, cultural, and historical context. He does not commit the mistake of regarding morality as a purely rational, transcendent, *a priori* enterprise by approaching it with a conception of philosophy that Putnam has called a "Master Science": "a discipline which surveys the special activities of natural science, law, literature, morality, etc., and *explains* them all in terms of a privileged ontology or epistemology that has proven to be an empty dream."[31] Finally, Kekes' account of the justification of moral intuitions requires participation in a situation. Because the assessment of an intuition is a matter of interpretation, it can be done only by those "who are involved in or witness the relevant situation." That is important for practical ethics because it imposes a significant constraint on the contribution of "armchair ethics."

The Nature of Practical Ethics

Moral problems in the real world, as opposed to the carefully pruned hypothetical examples of philosophers, are complex and murky, dynamic and evolving.[32] Two of the main tasks that confront those engaged in practical ethics are the identification and definition of moral problems and the description and individuation of the relevant facts and values. Both require familiarity with the general moral tradition within which one is operating and the specific moral tradition endemic to a clinical setting. Decisions also require familiarity with and understanding of moral traditions. Noble, for example, says that "the moral dimensions of conduct...are good or rational only taken as part of the realms of life they are born in."[33] And as she recognizes, assessments of moral traditions require empirical investigations:

> ...only...[the concrete] study [of social institutions] holds out hope of improving our understanding of moral standards, how they in fact operate, originate, and generalize themselves from one part of life to another, and thus how they may be improved, reformed, or better served.[34]

These tasks are foreign to philosophical ethics. A conception of morality in which it is exclusively rational and *a priori* ignores the difficulties surrounding how moral problems are seen and how moral traditions are applied and assessed. Yet these are the problems that by and large comprise practical ethics.[35]

Practical ethics is a combination of descriptive and critical ethics. Descriptive ethics is a necessary propædeutic to critical ethics. It is essentially a matter of understanding the moral tradition of a clinical setting—of understanding how facts are interpreted and why they are interpreted in this way. It requires being in the clinical setting and observing what goes on. Reading a book about, say, neonatal intensive care units, is not a substitute because although it might explain that certain selected facts are deemed important and some of the ways in which these facts are taken to be important, it cannot provide a comprehensive account of what is and what is not important, and why it is important, because it cannot capture the more recondite, tacit dimensions of a moral tradition. More important, reading about the moral practice of a unit does not permit an independent identification of facts that within the unit are *not* seen as morally charged, but might appear morally charged to an outsider. Yet when a moral tradition is being assessed, the facts that a tradition ignores can be as important as the facts that it recognizes.

One who remains skeptical might try the following experiment. Read about and examine the photograph of a painting in a book. Then take the book to a museum and read it in front of the painting. Is one's understanding of the painting enhanced by this direct confrontation? The point can also be made by exposing a failing in traditional analytic epistemology's preoccupation with knowledge. This approach to epistemology cannot explain how two persons with the same knowledge can have different understandings, that is, how one person could have a deeper understanding than another. Clinical experience is a way of acquiring understanding, not just knowledge, and that understanding is the basis of creative problem solving, regardless of whether the problem is moral.[36]

The critical side of practical ethics has at least four components. The first concerns the identification of moral problems. An outsider in a clinical setting brings a new way of seeing to the setting and thus can spot moral problems that those working in the setting might miss. Those in a unit might fail to recognize new facts as morally relevant or to assign new interpretations to facts because they are accustomed to seeing the same facts in the same way. Caplan's example of how medical personnel did not regard interviewing an elderly woman while she was defecating as an invasion of her privacy provides a nice illustration.[37]

The second involves the application of a moral tradition. An outsider who understands the tradition can check that the set of facts comprising a situation is properly identified. An outsider also can serve as a check on the self-centeredness of those who work in the clinical setting and thereby ensure that appropriate significance is attached to facts. In addition, an outsider can prevent contexts from being confused by, for example, pointing out that a genuine moral conflict exists, so appeals to the intuitions formed by a moral tradition are out of place. In other words, an outsider can remind those in the unit that moral conflicts are not to be resolved by appeals to the moral tradition of the unit.

The third contribution pertains to the handling of moral conflicts. Practical ethics can deal with moral conflicts through creative dissolution or creative resolution. Creative dissolution proceeds by discovering a new way of conceiving a problem or a new way of thinking about a problem that eliminates the conflict. Caplan's suggestion that air conditioners be provided to patients with respiratory ailments removed the need to allocate a hospital emergency room's scarce supply of oxygen units, for example.[38] As Caplan notes, in this instance and in the instance of getting medical personnel to accord more respect to a patient's privacy, effective moral action was "hardly dependent on analytical rigor or theoretical moral sophistication....Indeed, in both cases, ethical theory would have been the wrong place to turn for a solution to the issues...."[39] Creative resolution can occur when conflicting moral considerations are viewed as values rather than principles. A principled approach drives one to choose between competing principles. One must opt for either the principle of autonomy or the principle of beneficence, for example. Values lack this all-or-nothing dimension and thus permit the possibility of a compromise that maximizes fidelity to the competing values.

Finally, many people would find an approach to practical ethics that relies on the notion of a moral tradition too subjectivistic or relativistic unless moral traditions can be critically assessed. How is such assessment possible? One contribution that an outsider can make is determining how well the moral tradition of a clinical setting comports with the moral tradition of society. But something more than comparative assessments are necessary. Moral traditions themselves must be subject to critical appraisal. On what grounds can that occur?

Moral traditions can be criticized along the same lines that Geuss suggests ideologies can be criticized.[40] Geuss holds that a form of consciousness is ideologically false if it commits an epistemic, functional, or genetic mistake. These three categories of mistakes will be considered in order.

1. There are four kinds of epistemic mistakes. One type is mistakes about epistemic status. In biomedical ethics, an example would be thinking that additional scientific evidence will solve the issue of when the fetus becomes a person. A second is a false belief to the effect that some social

phenomenon is a natural phenomenon. An example is the belief that a woman's place is in the home because so many women, as a matter of fact, spend so much time doing housework. A third is a false belief to the effect that the particular interest of a subgroup represents the general interest of a group as a whole. An example is the belief that what is in the interest of doctors also is in the interest of society. The fourth is mistaking self-fulfilling beliefs for beliefs that are not self-fulfilling. Two examples are beliefs concerning the capacities of mentally handicapped persons and beliefs about the abilities of patients to handle bad news.

2. Geuss' notion of functional criticism cannot be directly transposed to moral traditions because ideologies are taken to be world pictures that stabilize or legitimize domination or oppression. But functional criticism nevertheless is apposite because moral traditions have the function of providing external and internal goods to those within a tradition. Insofar as a moral tradition fails to secure or distribute these goods, then, the tradition is subject to criticism and reform.

3. Because they accept the dogma of the genetic fallacy, philosophers dismiss considerations pertaining to the origin, motivation, or causal history of a belief as relevant to critical appraisal of that belief. But as Geuss points out, there are two senses in which genetic factors are relevant to the assessment of moral beliefs or standards. First, discovering the reasons why one in fact holds a belief and seeing whether one can acknowledge those reasons can be part of the assessment process. Second, understanding the purpose of a moral standard can be derived from an understanding of its history. The obvious parallel here is the interpretation of legal statutes. One guide to interpreting statutes involves an appeal to the end, goal, or purpose of the statute, for example, determining the particular defect that enactment

of the statute was intended to remedy. These genetic considerations obviously presuppose that moral beliefs and standards have causal histories and purposes, that they are not a set of timeless truths, but that presupposition is satisfied by the elements that comprise a moral tradition.[41]

These are examples of the types of rational criticisms to which moral traditions are susceptible. The examples display two distinctive features. First, they are directed to moral beliefs and standards that are located in a particular temporal and social context. They assume that morality evolves and that rational criticism can contribute to moral change. Second, these piecemeal criticisms are consistent with the weak rationality exhibited in morality. They amount to providing reasons for or against moral beliefs and standards, not constructing a rational foundation for the corpus of morality in terms of which every moral issue can, at least in principle, be uniquely settled.

Replies to Objections

A number of objections might be lodged against this account of practical ethics.[42]

Objection 1. Practical ethics is conflict resolution, not determining what is ethically obligatory, permissible, or wrong.

Reply. This objection begs the question. It presupposes that the "correct" conception of morality is provided by philosophical ethics. The content of morality is assumed to be equivalent to some philosophical moral theory, and the epistemology of morality is assumed to be some version of foundationalism. That view simply reflects the prejudice of philosophers. One would think that the extent to which philosophical ethics ignores "real life" morality would count more strongly against it, even in the eyes of philosophers.

Objection 2. At its worst, practical ethics takes as relevant whatever desires, aims, and concerns people may have, no matter how irrational or immoral; at best, it evaluates them on unstated, vague, or ad hoc grounds.

Reply. Not true. Relevancy in practical ethics is determined by a moral tradition, which itself is amenable to rational criticism. So the criteria of relevancy in practical ethics are not unstated. In addition, these criteria are certainly less vague than any putative criteria provided by a philosophical theory, such as the "utilities" of individuals or respect for persons. And to regard these criteria as ad hoc again begs the question because it presupposes that the only non ad hoc criteria are ones that flow from a philosophical theory.

Objection 3. Practical ethics may lack a requirement of consistency.

Reply. The strength of the consistency requirement in philosophical ethics is nothing to brag about. The requirement that like cases be treated alike presupposes an adequate account of relevant similarities, and philosophical ethics has never produced one. Moreover the admired, but elusive consistency of philosophical ethics seems to presuppose that uniquely correct answers to moral problems exist. Given the same relevant facts and moral considerations, morality demands that different persons arrive at the same decision. Practical ethics, on the other hand, leaves open the possibility that two people might arrive at different, but nonetheless morally permissible, outcomes. Confronted by the same facts and conflicting values, the more creative problem solver might come up with a better solution. Is the requirement of "universalizability" intended to preclude this possibility?

Objection 4. The consistency that results in practical ethics is unsatisfactory because it lacks an adequate criterion of relevant similarities.

Reply. *Tu quoque.* See the reply to objection 3. Again, the criteria of relevant similarity are provided by a moral tradition and are more likely to be helpful, because they are more specific and determinate than any criteria provided by moral theories.

Objection 5. Practical ethics has not produced any more specific or useful results than philosophical ethics, so how is it an improvement?

Reply. I suspect that this criticism is directed at the formulation of social policies with respect to moral matters, rather than at the resolution of concrete moral problems. Even at that level, however, the nod must go to practical ethics. The work of groups such as the National Commission for the Protection of Human Subjects of Biomedical and Behavioral Research in the US and the Committee of Inquiry into Human Fertilisation and Embryology in England is more compatible with practical ethics than philosophical ethics for the reason noted by Toulmin: such groups find it easier to reach a consensus about how specific types of problematic cases should be handled than to justify their decisions in terms of general views and abstract principles.[43] Toulmin reports that members of the National Commission "showed a far greater certitude about particular cases than they ever achieved about general matters."[44] Moreover, if one were to study the handling of concrete moral problems, I have no doubt that practical ethics would provide a better account of what actually occurs and what should occur.[45] Good examples of actual moral problem solving are usually cases of creative dissolution or compromise, and both notions are antithetical to philosophical ethics.

Conclusion

Rather than the extreme thesis I have been advocating, why not concede the possibility of compatibilism? Why not recognize that philosophical ethics and practical ethics simply operate at different levels? Philosophical ethics is concerned with everything above the level of rules or principles in the four versions of foundationalism I outlined, whereas practical ethics is concerned with everything below the level of rules or principles. Bayles suggests this kind of reconciliation:

> People in everyday life appear to use mid-level principles, otherwise few practical justifications would ever occur. Only moral theorists try to give complete, logical justifications. Mid-level principles at least partially explain why moral theorists make such little difference in the practical world; their differences are simply often not practically relevant.[46]

This suggestion misconceives the radical nature of the attack on philosophical ethics. Three points need to be emphasized. First, philosophical ethics and the foundationalism it embodies proceed from a conception of philosophy as a privileged discipline that Putnam has called "a Master Science." This conception is under attack on a number of fronts, but there is no realm in which real-world phenomena should be more relevant to philosophizing than morality. Philosophical ethics is objectionable because it imposes its view of what morality is on the real world, rather than constructing an account of morality that is informed by the real world. As Noble has observed, rather than broadening its compass to capture the phenomena of ordinary moral reflection, philosophical ethics narrows the ambit of moral reflection until it coincides with what is possible with moral theory.[47]

One genuine worry should be deflected. This attack on philosophy as "a Master Science" does not require that rationality be abandoned, only that its claims become more modest. The more ambitious pretensions of rationality have never been realized anyway in morality, law, or anywhere else, so the attempt of practical ethics to determine the boundaries of moral rationality should be welcomed.

Second, even in its own terms the compatibilist proposal fails. Bayles acknowledges, for example, that the same mid-level principle can be derived from different moral theories. If so, what real work are moral theories doing? Why are they not, as Noble charges, simply "ceremonial?"[48] Why should philosophical ethics not be regarded merely as an elitist activity for those who have the time and the inclination and, perhaps, get paid for it?

Finally, and most important, philosophical ethics can be dangerous. Approaching problems with a moral theory in hand can distort the subject matter of biomedical ethics.[49] The concern becomes spelling out the implications of that theory rather than trying to understand the nature of the conflict that generated the problem and searching for creative resolutions of the conflict.

Lest I lose my card-carrying privileges among philosophers, let me acknowledge, in conclusion, that the conceptual analyses that constitute "ground preparation" in an area can be a valuable contribution to practical ethics. They provide one kind of rational appraisal of a moral tradition. I am not opposed to this type of philosophical theorizing, only to the role that philosophical theories are taken to play in practical ethics. But if philosophers want to do more than prepare the ground, if they want also to harvest something from the ground, they need to get their hands dirty in the fields.

References

[1]My anecdotal, impressionistic experience suggests that allegiance to philosophical or practical ethics corresponds roughly to the absence or presence, respectively, of clinical experience on the part of those who hold these positions. In this respect the debate also concerns the role of clinical experience in biomedical ethics.

[2]Paul Ramsey (1976) "Conceptual Foundations for an Ethics of Medical Care: A Response," *Ethics and Health Policy*, Robert Veatch and Roy Branson (eds.), Ballinger Publishing Co., Cambridge, MA, p. 35.

[3]Caplan calls this conception of applied ethics the "engineering model." *See*, Arthur L. Caplan (1980) "Ethical Engineers Need Not Apply: The State of Applied Ethics Today," *Science, Technology, and Human Values* 6, 24–32; (1982) "Mechanics on Duty: The Limitations of a Technical Definition of Moral Expertise for Work in Applied Ethics," *Canadian Journal of Philosophy* VIII (Supp. vol.), 1–18; (1981) "Applying Morality to Advances in Biomedicine: Can and Should This be Done?" *New Knowledge in the Biomedical Sciences*, William B. Bondeson, H. Tristram Engelhardt, Jr., Stuart F. Spicker, and Joseph M. White, Jr., (eds)., D. Reidel, Dordrecht, pp. 155–168; and (1983) "Can Applied Ethics Be Effective in Health Care and Should It Strive to Be?" *Ethics* 93, 311–319.

[4]Terrence F. Ackerman (1980) "What Bioethics Should Be," *Journal of Medicine and Philosophy* 5, 274.

[5]Cheryl N. Noble (1979) "Normative Ethical Theories," *The Monist* 62, 496. *See also*, her "Ethics and Experts" (June, 1982) *Hastings Center Report* 12, 7–9.

[6]These are derived from Michael Bayles (1984) "Moral Theory and Application," *Social Theory and Practice* 10, 97–120 and (1986) "Mid-Level Principles and Justification," *Justification* (Nomos XXVIII), J. Roland Pennock and John W. Chapman (eds.), New York University Press, New York, NY, pp. 49–67.

[7]Noble, *supra*, note 5, stresses this point.

[8]Mary Hesse (1973) "In Defence of Objectivity," *Proceedings of the British Academy*, Oxford University Press, London, p. 8.

[9]Martin P. Golding (1984) *Legal Reasoning*, Knopf, New York, NY, pp. 8, 64.

[10]Joel Feinberg (1986) "Wrongful Life and the Counterfactual Element in Harming," *Social Philosophy and Policy* **4**, 163,164.

[11]The perduring attraction of strong rationality perhaps derives from the seemingly "scientific" nature of its rhetoric and its mode of thought. Fletcher suggests that this is why "reasonableness" questions are so alluring to lawyers. *See,* George P. Fletcher (1972) "Fairness and Utility in Tort Theory," *Harvard Law Review* **85,** 571–573.

[12]*See,* for example, Carol Gilligan (1982) *In a Different Voice,* Harvard University Press, Cambridge, MA.

[13]The full text is in section 24 of the *Canadian Charter of Rights and Freedoms,* in the *Constitution Act, 1982,* enacted by the *Canada Act, 1982,* c. 11, Schedule B (UK).

[14]For a discussion of these principles, *see* Tom L. Beauchamp and James F. Childress (1979) *Principles of Biomedical Ethics,* Oxford University Press, New York, NY.

[15]H. L. A. Hart (1961) *The Concept of Law,* Oxford University Press, Oxford, p. 125.

[16]Gidon Gottlieb (1968) *The Logic of Choice,* Macmillan, New York, NY, p. 56.

[17]*Ibid.,* p. 57.

[18]*See,* for example, Florence Kluckhohn and Fred Strodtbeck (1961) *Variations in Value Orientations,* Row Peterson, Westport, CT.

[19]I am indebted to Dr. George Deagle for this example.

[20]I owe this example to Professor Bernard Dickens.

[21]*See,* "Is Abortion the Issue?" (July, 1986) *Harper's* **273**, 35–43.

[22]Hilary Putnam (1983) "How Not to Solve Ethical Problems" The Lindley Lecture, Department of Philosophy, University of Kansas, p. 3.

[23]*Ibid.,* pp. 3–4.

[24]John Kekes (1986) "Moral Intuition," *American Philosophical Quarterly* **23,** 83–93. *See also,* his "Moral Conventionalism," (1985) *American Philosophical Quarterly* **22,** 37–46.

[25]For a more detailed explanation and defense of this notion of moral intuitions, *see,* Stuart Hampshire (1983) *Morality and Conflict,* Harvard University Press, Cambridge, MA, pp. 13–17.

[26]This clarification was suggested to me by Glenn Pearce.

[27]Kekes, "Moral Intuition," *supra,* note 24, p. 92.

[28]*Ibid.,* p. 89.

[29]The justification of moral traditions does not give up the game to philosophical ethics by covertly introducing rule utilitarianism into this account of morality. Kekes unfortunately suggests this criticism in his remark about "as favorable balance of benefit over harm for its members

as possible." But a weaker claim that omits the notion of maximization is possible and preferable. A moral tradition is justified only if it provides a minimum or threshold level of external and internal goods to its members. The justification is purposive, not utilitarian. Morality is a social institution that has a function, and both justification and critical appraisal proceed in terms of whether and how well it performs this function. In addition, the relative, conventional nature of internal goods is inconsistent with at least some forms of rule utilitarianism.

[30]This is Hart's argument about the *minimum* content of natural law. Hart, *supra*, note 15, pp. 189–195. Again, it is this emphasis on the minimum rather than the maximum that prevents Kekes' view from collapsing into rule utilitarianism.

[31]Putnam, *supra*, note 22, p. 8.

[32]Caplan, "Can Applied Ethics Be Effective in Health Care and Should It Strive to Be?" *supra*, note 3, p. 316.

[33]Noble, "Normative Ethical Theories," *supra*, note 5, p. 498.

[34]*Ibid.*, p. 500.

[35]For a discussion of these problems in the context of family medicine, *see*, Paul Rainsberry (1985) "The Responsible Family Physician: Rule Bound or Risk-Taking?" *Canadian Family Physician* 31, 439–445.

[36]I owe the points in this paragraph to Patrick Maynard.

[37]Caplan, "Can Applied Ethics Be Effective in Health Care and Should it Strive to Be?" *supra*, note 3, pp. 311–312.

[38]*Ibid.*, p. 312.

[39]*Ibid.*

[40]Raymond Geuss (1981) *The Idea of a Critical Theory,* Cambridge University Press, Cambridge, UK, pp. 26–44. I am indebted to Alison Wylie for calling this material to my attention.

[41]In other words, the genetic fallacy is viewed as a fallacy in morality because morality is taken to be more like mathematics than law. This again reflects the prejudice of philosophers in regarding the epistemology of mathematics as the "gold standard" and trying to force morality to "live up" to this standard. Because there is a proof procedure in mathematics, distinguishing the context of discovery from the context of justification is important. But in areas such as morality and law where there is no analogous proof procedure, the distinction breaks down. When this privileged, but inappropriate, epistemology is transferred to morality, it produces a backwards methodology. Instead of putting epistemology before content, content should come first in morality. Morality should be investigated in its own right and an epistemology appropriate to it developed from this investigation.

[42]These objections are culled from Bayles, "Moral Theory and Application," *supra,* note 6, pp. 99–102.

[43]Stephen Toulmin (1982) "How Medicine Saved the Life of Ethics," *Perspectives in Biology and Medicine* **25,** 741, 742.

[44]*Ibid.,* p. 742.

[45]For a distinctly nonphilosophical ethics account of how parents actually make decisions about whether to procreate after genetic counseling, *see,* Abby Lippman-Hand and F. Clarke Fraser (1979) "Genetic Counseling: Parents" Responses to Uncertainty," *Birth Defects: Original Article Series* **XV,** 5C, 325–339.

[46]Bayles, "Mid-Level Principles and Justification," *supra,* note 6, p. 65.

[47]Noble, "Ethics and Experts," *supra,* note 5, p. 9.

[48]Noble, "Normative Ethical Theories," *supra,* note 5, p. 506.

[49]Ackerman, *supra,* note 4, p. 271.

INDEX